23833151
3-31-08

Mathematical Knowledge

Mathematical Knowledge

Edited by
Mary Leng
Alexander Paseau
and Michael Potter

OXFORD
UNIVERSITY PRESS

Great Clarendon Street, Oxford OX2 6DP

Oxford University Press is a department of the University of Oxford.
It furthers the University's objective of excellence in research, scholarship, and education by publishing worldwide in

Oxford New York

Auckland Cape Town Dar es Salaam Hong Kong Karachi
Kuala Lumpur Madrid Melbourne Mexico City Nairobi
New Delhi Shanghai Taipei Toronto

With offices in

Argentina Austria Brazil Chile Czech Republic France Greece
Guatelmala Hungary Italy Japan Poland Portugal Singapore
South Korea Switzerland Thailand Turkey Ukraine Vietnam

Oxford is a registered trade mark of Oxford University Press
in the UK and in certain other countries

Published in the United States
by Oxford University Press Inc., New York

© The Several Contributors 2007

The moral rights of the authors have been asserted
Database right Oxford University Press (maker)

First published 2007

All rights reserved. No part of this publication may be reproduced, stored in a retrieval system, or transmitted, in any form or by any means, without the prior permission in writing of Oxford University Press, or as expressly permitted by law, or under terms agreed with the appropriate reprographics rights organizations. Enquiries concerning reproduction outside the scope of the above should be sent to the Rights Department, Oxford University Press, at the address above

You must not circulate this book in any other binding or cover
and you must impose the same condition on any acquirer

British Library Cataloguing in Publication Data
Data available

Library of Congress Cataloging in Publication Data
Data available

Printed in Great Britain
on acid-free paper by
Biddles Ltd, King's Lynn, Norfolk

ISBN 978-0-19-922824-9

10 9 8 7 6 5 4 3 2 1

QA
8
.4
.M385
2007

Preface

This volume stems from a conference on *Mathematical Knowledge* held in Cambridge from 30 June to 2 July 2004. All but one of the chapters (the exception being Cappelletti and Giardino's) were presented there, in some form or other. When we organized the conference, we intended it to start a dialogue between mathematicians, philosophers, and psychologists about mathematical knowledge, what it is and how it arises. As it turned out, the conference was also well attended by people from other disciplines such as mathematics education and computer science. Some of the speakers who helped to make the conference a success (Brian Butterworth, Susan Carey, Aki Kanamori, Charles Parsons, Gideon Rosen) have, for different reasons, not been able to include their contributions in this volume. However, what is here is more than enough to give the reader a lively sense of the the current state of the various debates. We are grateful to all the participants, to Adam Stewart-Wallace for compiling the index, but especially to the Jesus College Science and Human Dimension Project (through its trustee John Cornwell) and the Analysis Trust for the financial support that made the conference possible.

June 2007 M.C.L., A.C.P., M.D.P.

List of contributors

Alan Baker received his Ph.D. from Princeton University and was a Junior Research Fellow at Wolfson College, Cambridge. He is now an Assistant Professor in philosophy at Swarthmore College.

Marinella Cappelletti is at the Institute of Cognitive Neuroscience at University College London. Her current research investigates acquired disorders of number processing and degradation of numerical knowledge in patients with degenerative disorders.

Mark Colyvan is Professor of Philosophy at the University of Sydney. He is the author of *The Indispensability of Mathematics* (2001) and co-author of *Ecological Orbits: How planets move and populations grow* (2004).

Valeria Giardino obtained her Doctorate in logic and epistemology at the Faculty of Philosophy, University of Rome 'La Sapienza'. She is particularly interested in philosophy of mathematics and its possible connections with cognitive science. She is currently working on the use of diagrams in reasoning.

Timothy Gowers FRS is the Rouse Ball Professor of Mathematics at Cambridge University and a Fellow of Trinity College. He is the joint editor of the *Princeton Companion to Mathematics* (2007) and published *Mathematics: A very short introduction* in 2002. He was awarded the Fields Medal in 1998 for his research on functional analysis and combinatorics.

Mary Leng is a Lecturer in philosophy at Liverpool University. She was previously a Research Fellow at St John's College, Cambridge, and has held visiting fellowships at the Peter Wall Institute for Advanced Studies at the University of British Columbia and the Department of Logic and Philosophy of Science at the University of California at Irvine. Her book, *Mathematics and Reality* is soon to be published by OUP.

Alexander Paseau is a University Lecturer in philosophy at Oxford University and the Stuart Hampshire Fellow at Wadham College. He was previously a Research Fellow at Jesus College, Cambridge.

Michael Potter is Reader in the Philosophy of Mathematics at Cambridge University and a Fellow of Fitzwilliam College. His books include *Reason's Nearest Kin* (2000) and *Set Theory and its Philosophy* (2004).

Crispin Wright FBA is Wardlaw Professor of Logic and Metaphysics at St Andrews University. His books include *Frege's Conception of Numbers as Objects* (1984) and (with Bob Hale) *The Reason's Proper Study: Essays towards a neo-Fregean philosophy of mathematics* (OUP, 2001).

Contents

Introduction

MARY LENG

What is a number, that a man may know it, and a man, that he may know a number? (Warren McCulloch, *Embodiments of Mind* (1965))

Though posed by a neurophysiologist, McCulloch's question succinctly expresses the central epistemological problem of the philosophy of mathematics. Indeed, broadened slightly, the question 'What is mathematics, that humans may know it, and what is a human, that humans may know mathematics?' brings us straight to the main difficulty facing contemporary philosophical accounts of mathematical knowledge. There are, of course, many potential 'problems' of mathematical knowledge, and there is room for many fruitful approaches to these problems from a variety of perspectives and a variety of disciplines. But for the purposes of this Introduction the focus will primarily be on introducing what is sometimes regarded as *the* philosophical problem of mathematical knowledge, and considering how what philosophers may learn from psychologists and mathematicians in approaching this problem.

1 *Benacerraf's worry and some potential solutions*

The central philosophical problem of mathematical knowledge was articulated in its most well-known form by the philosopher Paul Benacerraf (1973).[1] His worry is simply this: the most obvious answers to the two questions 'What is a human?' and 'What is mathematics?' together seem to conspire to make human mathematical knowledge impossible. After all, we, as humans, are spatiotemporally located, embodied creatures. Our knowledge of the world we inhabit is mediated by our senses, and, one might assume, is therefore limited to what can justifiably be inferred on the basis of our sensory experience. Not only this but (in the light of worries first raised by the philosopher Edmund Gettier (1963) regarding the possibility of justified, true beliefs that ought not to count as knowledge) it is commonly supposed that, in order for a belief that p to count as a genuine case of *knowledge*, there must be some suitable connection between the *fact* that p and the grounds on which we have inferred that p is true.

[1]Michael Potter in his contribution to this volume describes it as a 'painful cliché' to begin in this way, but precisely because Benacerraf's problem has been so dominant in shaping recent philosophical writing, it seems appropriate in this Introduction to succumb once more to the cliché.

To illustrate why this condition is imposed, consider the plot of *The Importance of Being Earnest*, which rests on a Gettier-style situation. Gwendolen (quite reasonably) believes that Jack's real name is 'Ernest', for he has introduced himself as Ernest to her and goes by that name amongst his city friends. And Gwendolen's belief is, as a matter of fact, *true*, since, as it turns out, Jack was indeed christened 'Ernest' before being lost as a baby. But we would not wish to count Gwendolen's belief (that her suitor's real name is 'Ernest') as a genuine case of *knowledge*, even though it is both justified and true, since the fact that Jack's real name is 'Ernest' is entirely unrelated to the reason Gwendolen has for believing it to be so. (Indeed, she would have still believed him to be an Ernest even if his real name was Jack.) For us to *know* that p, then, it is not usually enough that we have good reasons to believe that p is true; our reasons for believing that p must appropriately track whatever it is that makes p true. Knowledge cannot simply be a matter of good luck.

But if we turn to the question 'What is mathematics?', the prospects for establishing such a link between the mathematical 'facts' and the justifications we standardly give for our mathematical beliefs—a link which would show that these beliefs are more than just lucky guesses—may seem slight. For many of our most firmly held mathematical beliefs appear to concern how things are with a realm of non-spatiotemporal, mind-independent objects (the kinds of things that philosophers would call *abstract*). For example, Euclid's proof that there are infinitely many prime numbers is not about the existence of concrete, spatiotemporal objects: the physical world might turn out only to be finite, and yet Euclid's proof of the infinity of primes would still be considered by most to be watertight. Neither, it appears, is Euclid's result a truth about objects that exist only in our minds (as ideas); for, once more, even if our minds turned out to be finite (as, indeed, they probably are), this would not, most agree, invalidate the claim that there are infinitely many primes. But Euclid's result does appear to be about *objects*, the prime numbers, and so it would appear that its truth rests on how things are with abstract objects, objects whose existence is independent of space, time, and human minds. And if we insist on imposing the condition that our reasons to believe claims about these objects must appropriately track how things *are* with such objects, we will find ourselves in some difficulty. For it is entirely unclear by what mechanism the justifications available to us (as world-bound, spatiotemporal creatures) for believing mathematical claims could appropriately track how things *really are* with the abstract objects about whose nature our mathematical claims appear to pass judgement. So it becomes unclear how genuine mathematical knowledge (of, for example, the infinitude of the primes) is possible.

Hold on, though—isn't there something ridiculous about this worry? Surely the infinitude of the primes is one of the things that we can be *most* certain about (more certain than we can ever be of contingent matters of fact about the physical world around us). After all, we said that Euclid had *proved* that there were infinitely many primes. Isn't the existence of such a proof enough to license us in saying that we *know* their infinitude? Doesn't the fact that we can prove mathematical theorems show that sceptical worries about the impossibility of mathematical knowledge are unfounded? As mathematician Joel Spencer puts it,

> in mathematics you can really argue that this is as close to absolute truth as you can get. When Euclid showed that there were an infinite number of primes, *that's it*! There are an infinite number of primes, no ifs, ands, or buts! That's as close to absolute truth as I can see getting. (Quoted in Hersh 1997: 11)

With this kind of view in mind, many will conclude that the problem lies not so much with the claim that we know that there are infinitely many primes, but rather with the reasoning that suggests that our knowledge of such facts is impossible. Thus, standard responses to Benacerraf's worry about our ability to know mathematical truths have seen it as challenging us to show precisely which of his assumptions about truth, knowledge, and the nature of mathematical objects ought to be rejected. Since we clearly do know some mathematical truths (this line of argument goes), there must be something wrong with the assumptions that together led us to the conclusion that such knowledge is impossible.

But it is worth noticing that the sceptical denial that mathematical proof can provide justification for believing mathematical theorems to be true is not quite as ridiculous as it sounds. A sceptic about our ability to know truths about mathematical objects need not question the assumption that mathematical proof gives us knowledge of *some* sort. For such a sceptic may allow that, by proving a mathematical result on the basis of axioms and definitions, we can come to know (perhaps even with certainty) what *follows* from our mathematical axioms and definitions. Yet the problem remains, the sceptic thinks, of showing how it is that our *axioms* track how things are with mathematical reality. Of course, if I *hypothesize* that zero is a number, that every number has a unique successor, and so on, I will soon (with the help of some judiciously chosen definitions) be able to prove that there are infinitely many prime numbers. But I can be confident that the conclusion I have proved reflects how things are in mathematical 'reality' only to the extent that I am confident that my axioms reflect the mathematical 'matters of fact', and for this, no appeal to the certitude of *proof* can help me.

Such concerns about our ability to know the axioms of our mathematical theories make scepticism about mathematical knowledge a viable position even in the light of the apparent certitude of mathematical proof. I defend a sceptical position concerning mathematical knowledge in my own contribution to this volume, where I argue that *fictionalists* who deny that we can know truths about mathematical objects can still lay claim to some logical knowledge concerning what does and does not follow from our axioms. The contrary position, that logical knowledge is just not enough to account for all we know when we know mathematics, is expressed by Michael Potter, who holds that attempts to replace mathematical knowledge by logical knowledge (e.g. of the consistency of theories) simply postpones the problem, on the grounds that our ability to lay claim to logical knowledge of this sort will itself depend on our having some genuine mathematical knowledge.

At any rate, it should be clear that the problem of how (if at all) we *know* mathematical truths is not simply answered by appeal to mathematical proof from axioms. If we want to preserve our claims to mathematical knowledge, we need either to give some reason for thinking that our mathematical *axioms* reliably reflect mathematical reality, or to show that this reliabilist way of posing the problem is mistaken. Various options are available for one who does not wish to embrace scepticism (with its rejection of many of our common-sense beliefs about mathematics). For example, perhaps the worry about the ability of our mathematical beliefs to track truth depends on an overly empiricist characterization of human knowers. Certainly, if we think that all of our knowledge must be gained from experience of the physical world, this will lead to difficulties in accounting for mathematical knowledge. But perhaps it is our empiricist assumptions that are mistaken here; perhaps there is room for a priori knowledge (knowledge that does not depend on experience) after all. Indeed, one might think that the fact that we have mathematical knowledge should be taken to be a refutation of strict empiricist assumptions about the grounding of all knowledge in experience.

The view that we can know mathematical truths a priori has a long and respectable history, and receives expression in contemporary philosophy of mathematics in, for example, the neo-logicist programme defended by Crispin Wright. According to logicists, we can know mathematical truths a priori since these truths are *analytic*, or truths of logic. Wright, for example, claims that certain second-order *abstraction principles*, from which various branches of mathematics follow, are conceptual, or logical, truths. If this is right, then simply by grasping the concept of natural number (for example) we can come to know that there are objects satisfying that concept, so we can rule out a priori the possibility of a gap between our mathematical axioms and the reality they aim to describe. Naysayers, however, will worry that there is something fishy about proving the existence of objects from our concepts alone, seeing the logicist's argument for the existence of numbers as worryingly similar to the ontological argument for the existence of God.

If we set aside the possibility of a priori knowledge of mathematical truths, and stick with the contemporary empiricist view that mathematical knowledge must be somehow grounded in experience, there remain several routes to avoiding the sceptical conclusion that human knowledge of ordinary mathematical truths is impossible. One route would be to question the assumption that mathematical truths must be truths about abstract objects. We might think that our earlier worries about the finitude of concrete, spatiotemporal objects would be enough to defeat any attempt to interpret mathematical truths as truths about concrete, spatiotemporal, objects. However, this is to underestimate the ingenuity of philosophers! Penelope Maddy, for example, argues in her (1990) book, *Realism in Mathematics*, that at least some mathematical objects (the impure sets, that is, sets such as the set of eggs in the fridge, the set of the set of eggs in the fridge, and so on, whose transitive closures contain concrete, physical objects),[2] are spatiotem-

[2]The transitive closure of a set A is obtained from A by applying repeated applications of the union

porally located, using the principle that a set is located where its members are. On Maddy's view, we can have ordinary *perceptual* knowledge of some mathematical objects (the impure sets) via our causal interactions with their members. In defending this claim, Maddy borrows from the work of psychologists such as Piaget (1937) and Hebb (1949) to give plausibility to the claim that we may have the requisite perceptual machinery to perceive not just concrete objects, but also sets of those objects. If she is right—and, unfortunately, this is a big if: most commentators have balked at the idea that one's fridge might contain, not just the carton of eggs, but the set of them, and the set of the set of them, and so on up the hierarchy—then a closer look at the nature both of human perceivers and of the mathematical reality perceived brings the two closer together in such a way that human knowledge of mathematical objects might seem explicable after all.

Alternatively, if we do not want to interpret mathematical truths as truths about concrete objects, and yet still have worries about our ability to gain knowledge of abstract objects, we might question whether (despite surface appearances) mathematical truths are truths about objects at all. For example, Philip Kitcher (1984) has argued that mathematical statements should be understood as truths about (idealized) human collecting and segregating activities, and has argued (in an updated defence of the empiricism of J. S. Mill) that we learn about them through experience of these activities. Other attempts at reinterpretation have suggested that mathematical claims which appear to be committed to the existence of abstract objects (such as 'There are infinitely many prime numbers') are actually shorthand for claims that do not carry with them such commitment. Thus Hellman's (1989) modal structuralism, for example, reinterprets 'There are infinitely many prime numbers' as really just shorthand for the more complex modal claim, 'It is necessarily the case that, *if* the Peano axioms hold, then there are infinitely many prime numbers'.[3]

Many options seem open, then, for interpreting mathematical claims so that they are no longer seen as making assertions about problematic abstract objects. If we could show that such an interpretation of our mathematical language is plausible, then we would be able to explain our mathematical knowledge without having to suppose that such knowledge is knowledge of abstract mathematical objects. But is it reasonable to suppose that, despite surface appearances, the truth of a claim such as 'There are infinitely many prime numbers' does not really require the existence of infinitely many objects? Is it fair on mathematicians to interpret them as not really meaning what they appear to say when they make such utterances? Many mathematicians would be surprised, to say the least, to

operator: it is $\bigcup\{A, \cup A, \cup\cup A, \ldots\}$. The transitive closure of A will contain a physical object if some member of a member of ... a member of A contains a physical object.

[3]Alternative versions of structuralism hold that mathematical theorems should be interpreted as truths about the positions in abstract structures, holding that our ability to abstract structural properties of concrete systems shows that we can have knowledge of abstract *structures* (see, e.g., Shapiro 1997). However, despite bringing in abstract *structures* in interpreting mathematical theories, such accounts still take mathematical truths to be truths about abstract objects—the positions in those structures.

discover that what they take to be assertions of truths about prime numbers are really just assertions about what follows from the Peano axioms. Indeed, many would think that it is inappropriate for outsiders such as philosophers to try and tell mathematicians what it is they really mean by their mathematical assertions.

Finally, if we do take mathematical knowledge to be knowledge of truths about abstract objects, the possibility remains open for empiricists to question our assumption that there must be some specific link which explains how it can be that our beliefs about abstract mathematical objects reliably reflect mathematical reality. Following W. V. Quine, empiricists who are also *holists* reject the idea that *individual* beliefs (such as our belief that there are infinitely many primes) can count as knowledge only if our justifications for *those* beliefs are appropriately responsive to the objects (in this case, the prime numbers) about which they appear to make claims. On holistic views of confirmation, what gives us reason to believe that p is just that p forms a part of our best overall theory, that theory being the result of our best efforts to organize our experience. We should not ask in isolation, therefore, whether we have reason to believe a particular hypothesis that forms a part of a given theory, but only whether we have reason to believe the theory in which that hypothesis occurs as a whole. And this is simply a question of how well that theory has stood up to the test of experience. In Quine's memorable words, 'our statements about the external world face the tribunal of sense experience not individually but only as a corporate body' (1953: 41).

Since mathematical hypotheses appear to form an indispensable part of our best empirical theories (our mathematically stated scientific laws presuppose the truth of a great deal of mathematics), whatever confirmation those theories receive via their empirical successes extends on the holistic view to the mathematical parts of those theories. So theoretical hypotheses concerning mathematical objects receive just as much confirmation as hypotheses concerning unobservable physical objects such as electrons; the supposed abstractness of the mathematical objects hypothesized is simply irrelevant. Objections to the Quinean holist view have emphasized various disanalogies between the mathematical and non-mathematical hypotheses of our theories, and have suggested that, *contra* Quine, these disanalogies show that all confirmation is not equal. Mark Colyvan responds to some of these objections in this volume, pointing to cases where mathematical hypotheses do appear to play a very similar role in our theories as is played by hypotheses concerning (for example), unobservable physical objects.

Alexander Paseau's chapter develops another objection to the Quinean position by suggesting that mathematical truths may be scientifically confirmed even if no particular interpretation of these truths is confirmed. Science confirms the truth of 'There are infinitely many prime numbers', but not its abstract object interpretation *there are infinitely many abstract prime numbers*, nor Hellman's modal structuralist interpretation, nor any other. If this is right, a scientific defence of the possibility of human knowledge of *mathematical truths* need not amount to a defence of the possibility of human knowledge of *abstract objects*.

What should be clear from this survey is that philosophers are far from agreed about how best to solve the 'problem' of mathematical knowledge. But given that the problem, as I have presented it, involves reconciling our conflicting collections of intuitions about, first, the nature of mathematics, and second, the nature of human cognition, one might reasonably hope to find some insight into how to deal with it by comparing notes with mathematicians and psychologists. It is the job of philosophers to come up with a plausible story of the nature of mathematics, in such a way as to account for the phenomenon of doing mathematics and the uses to which it is put. But surely the best source of data regarding the subject matter philosophers are trying to account for is mathematicians, those who actually participate in the practice that philosophers are trying to account for. And similarly, while it is the job of philosophers to provide an account of the nature of mathematical knowledge that explains how we humans are able to form mathematical beliefs that we can reasonably count as knowledge, surely one of the best ways of approaching this issue is to look at how we do in fact come to form our mathematical beliefs. And who better to look to for an answer to this question than psychologists? The question that naturally arises, then, is: What can philosophy learn from mathematics and psychology about the nature of mathematical knowledge?

2 *Mathematical knowers and mathematical knowledge: What can mathematicians and psychologists teach us?*

According to at least some mathematicians and psychologists, the answer to this question is, 'Everything'. That is, once one looks closely at mathematical practice and/or at our best theoretical accounts of the mathematical brain, the nature of mathematical knowledge becomes obvious, and philosophical puzzles dissolve (leaving philosophers looking rather dullheaded even to have worried about such things in the first place). On the side of mathematics, for example, amongst those mathematicians who have cared enough to grapple with philosophical problems in the first place, one often finds expressions of exasperation concerning a perceived unwillingness within the philosophical community to see the answer that is staring them in the face.

Mathematician Reuben Hersh (1997), for instance, has suggested that philosophical worries about human ability to know truths about mathematical objects disappear immediately once one realizes that mathematical objects are neither mental nor physical, but *social*. In reaching this conclusion, Hersh considers the question, 'How do flesh-and-blood mathematicians acquire the knowledge of number?', and argues that 'to answer, you have to forget platonism, and look in the socio-cultural past and present, in the history of mathematics' (1997: 13). A proper look at mathematics shows that mathematical theories have a history, that they result in part from social requirements and concerns, and that their acceptance is bolstered by the agreement of communities. Why not, in that case, accept mathematical objects for what they are—socially constructed artefacts of our social-cultural-historical reality? It is, Hersh seems to think, only philosophical stubbornness that keeps this solution from reaching acceptance.

Well, not quite. One worry that philosophers have with this kind of argument is the move from the claim that our *theories* have a history and arise out of social pressures to the claim that the *objects they concern* are socially constructed artefacts. Does the analogous argument work with *empirical* theories? Our empirical scientific theories have a rich history, and we can certainly recognize the impact of social-cultural pressures on the development of those theories. But does this mean that their objects exist only in some social-cultural-historical reality? Just because there is a social-cultural-historical story to tell about the genesis of the modern day periodic table, for example, does this mean we have to conclude that *elements* such as oxygen and hydrogen exist only as social constructs? While there are some who would wish to draw this conclusion, those who find the move questionable in the case of empirical science should pause before accepting the same move when it is made in debates about the nature of mathematics.

Secondly, there is a worry about whether the claim that mathematical objects exist only as social-cultural-historical artefacts will be enough to account for our standard assumptions about the truth-values of mathematical claims. Consider Goldbach's Conjecture, one of the unsolved problems discussed in Alan Baker's contribution to this volume. Most mathematicians assume that this conjecture is either true or false: we just don't know which. But if what makes it true or false is a social fact, is it safe to assume that social reality is substantial enough to answer this question? For Goldbach's Conjecture to be false, there would have to exist (in social-cultural-historical reality) an even number greater than 2 that is not a sum of two prime numbers. But we have certainly not constructed such a number (at least, not one that we know of). Should we say, for now at least, that Goldbach's Conjecture is simply *true*, since there isn't (yet) a counterexample in social-cultural-historical reality as it stands at the moment? Baker suggests that there is a wide level of agreement amongst mathematicians concerning the probable truth of this conjecture. So if mathematical truth is determined by social-cultural-historical factors, one might think it would be appropriate for mathematicians to take advantage of the current consensus and simply *stipulate* that Goldbach's Conjecture is true. Of course, a sensible social-constructivist will not accept either of these options, and will instead try to argue that there is enough in our mathematical practices to pin down a standard model containing all the natural numbers, and hence all potential counterexamples to Goldbach's Conjecture. But the difficulty remains of justifying the assumption that social-cultural-historical reality is going to be so extensive (not to mention the assumption that this 'reality' is consistent).

A related worry concerns mathematical claims that many would like to hold would have been true *even if there were no society of mathematicians to contemplate their truth*. Thus, in a review of Davis and Hersh (1981) Martin Gardner objects to their 'social' account on the ground that

> when two dinosaurs met two dinosaurs there were four dinosaurs. In this prehistoric tableau '$2 + 2 = 4$' was accurately modelled by the beasts, even though they were too stupid to know it and even though no humans were there to observe it. (1981)

Hersh's response to Gardner's objection is illuminating. Hersh points out that the statement 'two dinosaurs plus two dinosaurs equals four dinosaurs' makes only *adjectival* use of the words 'two' and 'four'. Indeed, we can rewrite this statement so that the words 'two' and 'four' don't even appear. So, Hersh concludes, this statement is not about *numbers*, but just about *dinosaurs* (and similarly for other applications of arithmetic to discrete physical objects):

> If I say 'Two discrete, reasonably permanent, non-interacting objects collected with two others makes four such objects,' I'm telling part of what's meant by discrete, reasonably permanent non-interacting objects. *That is a statement in elementary physics.* (1997: 15, my emphasis)

On Hersh's view, then, it was *true* that 'Two dinosaurs plus two dinosaurs made four dinosaurs' even in prehistoric times, before human societies had been able to create the numbers two and four. But this is because this statement is not *really* a truth about numbers (which didn't exist before humans got around to inventing them), but only a truth about dinosaurs (which did exist before humans). Because this statement only uses numbers adjectivally, and not as nouns, its truth does not require the existence of numbers, and so it is not *really* a mathematical statement, but simply a 'statement in elementary physics', the kind of thing that could have been true even if there were know humans around to entertain its truth.

Here, though, philosophers will begin to worry. For, despite its initial plausibility, the distinction Hersh is relying on here between statements of physics and statements of mathematics is not as clear cut as he seems to think. Hersh suggests that the truths of physics, being about concrete, physical objects and their relations to each other, could be true timelessly, independent of our human social practices, whereas truths of mathematics, being about socially constructed objects, only become true once humans are around to entertain their truth. But, at least if we look at our ordinary physical theories, the dichotomy between mathematical and physical truths becomes less clear. Many of the laws of physics, while ultimately being 'about' physical objects and their behaviour, are expressed in terms of the relations between these objects and mathematical objects such as numbers, functions, and sets. If we wish to hold that such laws (or laws like them) were true even before we humans were around to contemplate them, then it appears that we will have to accept that the mathematical objects they appear to refer to also existed before humans did.

In the classic statement of this difficulty, Hilary Putnam considers Newton's Law of Universal Gravitation, which asserts that there is a force f_{ab} exerted by any body a on any other body b. The direction of the force f_{ab} towards a, and its magnitude F is given by:

$$F = \frac{gM_aM_b}{d^2}$$

where g is the universal constant, M_a is the mass of a, M_b is the mass of b, and d is the distance which separates a and b. (1979: 338)

This Law is talking about physical bodies and the forces between them, but what exactly does it say about them? It says 'just that bodies behave in such a way that the quotient of two numbers *associated* with the bodies is equal to a third number *associated* with the bodies' (1979: 74). The Law of Universal Gravitation is, therefore, as much a claim about mathematical objects (real numbers and 'associations'—i.e. functions) as it is a claim about bodies and the forces between them.

If we think that such laws are true (and were true even when the dinosaurs were around), this raises a difficulty for any attempt to say that mathematical objects depend for their existence on us. According to Hersh, we can accept the prehistoric truth of 'Two dinosaurs plus two dinosaurs makes four dinosaurs' only because we can show that this statement does not really make reference to numbers, but only to dinosaurs. But unless we can similarly show that the laws of physics in general do not really require us to talk about numbers or other mathematical objects, then we cannot persist in holding that the truth of physical laws depends only on the *physical* objects and systems of objects they concern. So if we do not think that numbers existed before we were around to invent them, neither can we think that the laws of physics were true even before we were around to discover them.

This is not to say that no options are open for the view of mathematical objects as social constructs: defenders of such a view could give up on the permanent truth of physical laws, holding that physical laws are just as socially dependent as mathematical statements are. Alternatively, they might follow Hartry Field (1980) in trying to show that the laws of physics *can* be rewritten so as to avoid any apparent reference to numbers. Or perhaps there is some further way of preserving the intuition that the laws of physics have some permanently true non-mathematical content, even if our best statements of those laws are mathematical (as argued in some recent defences of fictionalism, such as (Balaguer 1998)). At any rate, it should be clear that, even if we accept Hersh's point that mathematics occurs within a historical and social context, much philosophical work remains for anyone defending the view that mathematical reality is social reality.

Similar worries arise for some of the grand claims made by some psychologists who suggest that their work provides quick and easy answers to the traditional philosophical problems. For example, cognitive scientist Rafael E. Núñez recently teamed up with linguist George Lakoff to write an illuminating study of the development of mathematical concepts (2000). They have high hopes for the application of cognitive science to issues of mathematical knowledge.

It is *only* through cognitive science—the interdisciplinary study of mind, brain, and their relation—that we can answer the question: What is the nature of the only mathematics that human beings know or can know? (Lakoff and Núñez 2000: 3, my emphasis)

Why *only* through cognitive science? Because, they tell us,

The only conceptualization that we can have of mathematics is a human conceptualization. Therefore, mathematics as we know and teach it can only be humanly created and humanly conceptualized mathematics. (2000: 3)

Passages such as these, however, equivocate between the use of the word 'mathematics' to mean our mathematical *theories*, and to mean whatever it is that those theories are theories *about*. Granted, our mathematical theories are humanly created and humanly conceptualized, but this does not mean that they must be theories about humanly created objects. For it is equally true that physics as we know and teach it can only be humanly created and humanly conceptualized physics: the concepts in which we frame our theories are our own. But it does not follow from this that physics is a theory about our human concepts, so that we can only know the nature of the *subject matter* of physics through cognitive science. And even if we do think there are special reasons to take the subject matter of mathematics to reside in human minds, we will soon find ourselves encountering the same kinds of difficulties that plagued Hersh's account of mathematical objects as social constructs.

The discussion of these examples is not, though, meant to suggest that there is nothing to be learnt from mathematics and psychology about the nature of mathematical knowledge. On the contrary, as I will argue below, philosophers can expect to learn a lot from mathematicians and psychologists concerning these issues. What we should be cautious about, however, is the thought that a proper understanding of how mathematicians as a community have developed the mathematical theories that we find ourselves with, or of how human minds are set up so as to be able to grasp mathematical concepts and entertain mathematical beliefs in the first place, can be enough to answer *all* of the philosophical worries surrounding the issue of mathematical knowledge. For most would accept that there is a distinction to be drawn between our having mathematical beliefs and our having justification for those beliefs. A study of how, as a matter of fact, we have reached the mathematical beliefs we now accept, whether this proceeds via consideration of the history of mathematics or via an account of the brain-structures that have made us capable of mathematical thinking, can certainly tell us a lot about where our mathematical beliefs come from. But such a study need not even touch on the question of whether those beliefs are justified. In the light of even the most detailed account of how it is that we have the mathematical beliefs that we have, the question will remain of whether, and how, those beliefs are justified.

If we do recognize a separation between the question of how we have formed our mathematical beliefs and the question of whether beliefs formed in such a way can be thought of as justified, then it is perhaps best to see the testimony of mathematicians and the evidence provided by psychologists as an accompaniment to philosophical thought about mathematical knowledge, and not a replacement for it. And certainly, in this role, mathematical and psychological perspectives on our mathematical knowledge can be extremely valuable in guiding philosophical thought. If philosophers want to show that the mathematical beliefs we have have been reached via a route that is (given their particular account of the nature of mathematics) likely to lead us to mathematical knowledge, they will need to take care to ensure that their understanding of what mathematical beliefs we have, and how those beliefs have in fact been generated, is a sound one. Mathematicians' testimony and psychologists' empirical research thus places limits on what kinds of philosophical accounts are acceptable.

Take, for example, Kitcher's revival of Mill's empiricist account of mathematics, mentioned earlier. According to Kitcher, our most basic mathematical concepts are reached through our experience of engaging in various activities. 'Children', he tells us, 'come to learn the meanings of "set", "number", "addition" and to accept basic truths of arithmetic by engaging in *activities* of collecting and segregating' (1984: 107–8). Other mathematical concepts, he argues, have evolved historically out of successive levels of idealization from these basic ones. On the basis of this empirical hypothesis concerning the genesis of our mathematical beliefs, Kitcher suggests a hypothesis concerning the nature of mathematical truth that would explain why mathematical beliefs reached in this way would count as knowledge. 'Rather than interpreting these activities as an avenue to knowledge of abstract objects, we can think of the rudimentary arithmetical truths as true in virtue of the operations themselves.' (1984: 109) Kitcher's account of the nature of mathematics is thus guided by an empirical assumption about how our mathematical beliefs arise. It is attractive to the extent that, if this assumption about the genesis of our mathematical beliefs is true, it provides a story of how beliefs that we have come to in that way could plausibly be counted as knowledge.

But, as Marinella Cappelletti and Valeria Giardino point out in their survey in this volume of recent results in neuroscience relating to mathematical cognition, Kitcher's empirical hypothesis appears mistaken. For experiments such as Karen Wynn's famous (1992) study of arithmetical expectations in children as young as four months, suggest that some basic numerical abilities are present from birth, and that it is an innate capacity for recognizing numerosity, rather than learned generalizations concerning our ability to collect and segregate objects, that forms the cognitive basis for our most basic mathematical knowledge. While this does not refute Kitcher's empiricism as an account of the *justification* of our mathematical beliefs—perhaps our basic arithmetical beliefs, though innate, only receive genuine justification once they are shown to be grounded in true empirical generalizations about our collecting and segregating abilities—such studies do remove one of the main motivations for this kind of account of mathematics, namely that of basing our account of the nature of mathematics on an understanding of how we in fact come to our mathematical beliefs. Insofar as Kitcher's aim is to give an account of how the mechanisms by which we do *in fact* reach our mathematical beliefs can be thought of as reliable mechanisms (i.e. as reliably leading us to true mathematical beliefs), to that extent his failure to give a correct account of those mechanisms is fatal.

Kitcher's strategy to deal with the problem of mathematical knowledge was to start with an empirical hypothesis concerning how our mathematical beliefs are formed, and to provide an account of the nature of mathematics that would show that beliefs formed in that way count as knowledge. If we update the empirical hypothesis, then, an updated philosophical hypothesis might also emerge. Psychologists disagree about the precise nature of our innate capacities for arithmetical thinking. For example, at the conference from which this volume has emerged, psychologists Brian Butterworth and Susan Carey discussed their disagreement over the issue of whether the basic innate capacity that lead us to a concept of number is a capacity to grasp discrete *numerosities*, or to grasp differences in magnitude between *quantities*, which do not come neatly packaged into discrete single unit packages. Butterworth interprets the evidence from neuroscience as showing that we have an innate 'Number Module', which allows us to respond to the number of discrete elements in a collection (see his 1999). A competing hypothesis is defended by Stanislas Dehaene (1997), who argues for an 'accumulator model' of our innate ability to count, according to which our capacity to respond to, and have expectations about, the sizes of collections is mediated by an ability to weigh up and compare continuous quantities. While she thinks that small numerosities might well be dealt with as Butterworth suggests, via a specific number module, Carey suggests that Dehaene's model might be more plausible as an account of our facility with larger calculations.

If we wish to tie our philosophical account of mathematical knowledge closely to an account of how our mathematical beliefs have come about, differences such as these may matter: follow Butterworth, and we might be attracted to an updated version of Penelope Maddy's set-theoretic realism, holding that this evidence shows that we have the ability to perceive numerical properties of sets. Follow Dehaene and we might be more tempted to view the claims of arithmetic as abstractions and idealizations of more basic physical facts about differences in continuously changing quantities. At any rate, for those who wish to provide an account of the nature of numbers that explains why the routes by which we come to form beliefs about numbers are likely to be reliable, empirical studies of our capacities to form mathematical beliefs will be crucial.

How about mathematicians? What can philosophers learn from them about the nature of mathematical knowledge? One crucial point that mathematicians will wish to remind us of is that mathematics does not stop with basic arithmetic. If we wish to account for *mathematical* knowledge, we have to go beyond McCulloch's question, 'What is a number, that a man may know it?' Numbers are just part of the story, and any account of mathematical knowledge worth its salt will have something to say about the vast realm of mathematics beyond basic arithmetic. This is not to say that there is not something special about the counting numbers. Indeed, the studies of dual dissociations mentioned by Cappelletti and Giardino may be taken as evidence for our giving a special epistemic status to numbers, since it appears that processing numerical quantities is independent from other cognitive processes, including, presumably, those processes used elsewhere in mathematics. (There is anecdotal evidence for this difference too: many excellent mathematicians claim to be terrible at arithmetic.) We might, then, wish to agree with Kronecker that 'God created the natural numbers—everything else is the work of man.' But whether or not we accord a special status to the natural numbers, the point remains that if we wish to account for mathematical knowledge, we will need also to account for the sense in which we can be said to know the claims of mathematical theories outside of number theory.

Another thing that listening to mathematicians may teach us about the nature of mathematical knowledge is that, despite the understandable tendency amongst philosophers to focus on the epistemic status of axioms, proof from axioms is not the only means by which mathematicians claim to acquire mathematical knowledge. Indeed, if one takes seriously G. H. Hardy's perspective, the importance of proof as a route to mathematical knowledge becomes somewhat downgraded. 'I have,' Hardy tells us,

> always thought of a mathematician as in the first instance an *observer*, a man who gazes at a distant range of mountains and notes down his observations. ...If he wishes someone else to see it, he *points to it*, either directly or through the chain of summits which led him to recognize it himself. When his pupil also sees it, the research, the argument, the *proof* is finished.
>
> The analogy is a rough one, but I am sure that it is not altogether misleading. If we were to push it to its extreme we should be led to a rather paradoxical conclusion; that there is, strictly, no such thing as mathematical proof; that we can, in the last analysis, do nothing but point; proofs are what Littlewood and I call 'gas'; rhetorical flourishes designed to affect psychology, pictures on board in the lecture, devices to stimulate the imagination of pupils. This is plainly not the whole truth, but there is a good deal in it. (1929: 18)

If Hardy is right, a full account of the nature of mathematical knowledge will have to take into consideration potential routes to such knowledge aside from deductive proof. This challenge is taken seriously by Alan Baker, whose chapter in this volume considers the role of non-deductive evidence for mathematical beliefs, asking whether, in particular, mathematicians accept *enumerative induction* as a ground for believing mathematical propositions that are not yet proved.

Aside from drawing our attention to the varieties of justificatory methods at work in mathematical practice, mathematicians also remind us of the various kinds of mathematical knowledge. The problem of mathematical knowledge that we have primarily been discussing focuses on the issue of how we can know *that* the various claims that make up our mathematical theories are true, treating all kinds of mathematical 'knowledge-that' as on a par. But one message we can take from Timothy Gowers's contribution to this volume is that not all forms of mathematical knowledge are equal. Even if we just focus on mathematical statements for which we can find a proof, setting aside issues of alternative forms of mathematical justification, we quickly discover that some such statements are more interesting than others. After all, any one proof gives rise to infinitely many more trivial additional proofs: having proved that there are infinitely many prime numbers, we can deduce that 'either there are infinitely many prime numbers or $1 + 1 = 2$'; 'either there are infinitely many prime numbers or $1 + 1 = 3$'; and so on for as many disjuncts as we wish to entertain. But although all of these results count as mathematical knowledge, and are known as certainly as we know Euclid's theorem itself, proving them won't get anyone a Fields Medal, however many they succeed in proving!

Mathematicians, then, do not just care about proving theorems: they care about proving *interesting*, *deep*, *fruitful* theorems, by means of *elegant*, *ingenious*, *explanatory*, *memorable*, or even *amusing* proofs. If we wish to understand more about the character of mathematical knowledge, we ought to investigate these kinds of evaluative claims made by mathematicians. Gowers focuses on *memorability*as a desirable property of mathematical proof, and hypothesizes that what he calls the 'width' of a proof contributes greatly to its memorability and therefore its value. Gowers's emphasis on the value of memorability is reminiscent of Wittgenstein's concern that mathematical proofs should be *surveyable*. However, while for Gowers surveyability—or memorability—is to be preferred when we can get it, it is unlikely that as a working mathematician he would follow Wittgenstein in holding that *only* a 'structure whose reproduction is an easy task is called a "proof"' (1956: 143).

There is much, then, that can be learned from mathematicians and psychologists concerning the traditional philosophical questions about the nature of mathematical knowledge. An interdisciplinary approach can provide philosophers with new perspectives on old questions, as well as pointing to further questions that are worthy of philosophical attention. There remains, of course, the question of whether mathematicians and psychologists have anything to learn from *philosophers* about questions surrounding mathematical knowledge that are specific to their own disciplines. This is not a question that this Introduction, or indeed the volume as a whole, tries to answer. I do believe, however, that the philosophical issues which this collection discusses will be of interest to all those with an intellectual curiosity about the nature of mathematical knowledge.

What is the problem of mathematical knowledge?

MICHAEL POTTER

This is a book about the problem of mathematical knowledge. So it might be useful to try to get clear about what the problem is. Why do philosophers take there to be a particular problem, worth devoting a whole book to, about *mathematical* knowledge?

One thing I should say straightaway is that I do not intend anything I say here to be especially deep: if it is worth saying at all, that is only because some philosophers seem not to recognize it sufficiently. Another is that my aim here is modest: all I want to do is to try to get a bit clearer about what the problem is, not to propose a solution.

1 What is distinctive about the mathematical case?

With the preparatory throat-clearing out of the way, the first thing I think we need to get clear about is how the problem of mathematical knowledge differs from the problem of logical knowledge. One of the great failed projects in the history of the philosophy of mathematics was logicism, which aimed to show that mathematical knowledge was just a species of logical knowledge. If that project had succeeded, it would thereby have shown that there is no problem of mathematical knowledge, of how we come to know that mathematical theorems are true, distinct from the problem of logical knowledge, i.e. of how we come to know that arguments are logically valid.

But logicism did fail. Why it failed is a story too long to tell in any detail here. Briefly, it was because logic could not deliver a domain of objects, and a conception of the range of properties these objects may have, rich enough to support the whole of mathematics as standardly understood. Those for whom this brief summary is not enough should consult my book (2000) for more. What matters here is only to note that logicism's failure is the reason why there is still a problem of mathematical knowledge. What I want to do here, therefore, is to explore what this problem amounts to. I want to get clear, that is to say, about what *more* there is to the problem of mathematical knowledge than there is to that of logical knowledge.

By formulating the question in this way I do not want to be taken as claiming that there is not a problem of logical knowledge. There is. In the *Tractatus* Wittgenstein tried to advance the beautifully simple idea that there is no problem of logical knowledge because in logic there is, in a sense, nothing to be known. This, like logicism, would have been nice if it had worked. But it did not, which is why there is still a problem of logical knowledge.

Notice, though, that there was, nonetheless, something right about Wittgenstein's idea. There is little doubt that he under-estimated the complexity of logic. He thought that it is trivial, and in that he was wrong. But his reason for thinking it is trivial, namely that it says nothing about the world, surely has something right about it. It is a familiar experience of anyone who has taught modal logic that as soon as we mention 'possible worlds', some students will immediately assume that this is meant as a contrast with 'impossible worlds'. The point which Wittgenstein surely had right is that there is something fundamentally mistaken about the thought that there might be any sort of contrast to be had here: there is no such notion as that of a logically impossible world.[1]

2 *Implicationism all the way down?*

One of the complaints commonly made by mathematicians about the philosophy of mathematics is that it seems to concentrate so much on three special cases: arithmetic, analysis, and set theory. Don't philosophers realize, they ask despairingly, that most of modern mathematics is not about these but about abstract structures—groups, Hilbert spaces, infinite-dimensional manifolds, Lie algebras, and so on?

Well, yes they do realize this. The reason why they focus, and this chapter will focus, on the three special cases is not that they are the only cases philosophers know about (although there may no doubt be some philosophers of whom that is true) but because what makes these cases special is also what makes them specially problematic. If I prove something about all groups, or all Lie algebras, then what I know *is* just a piece of logical knowledge. The philosophical view known as implicationism (or, less elegantly, if-thenism) is a perfectly adequate explanation of what is going on in these cases.

Notice, though, that the theory of groups, or the theory of Lie algebras, is a matter not just of proofs but of counterexamples, and something more needs to be said about them. If we want to show that not all abelian groups are cyclic, what we do is to exhibit a non-cyclic group (the Klein four-group, for example). We cannot (at least without further explanation) treat this in an implicationist manner, since the conclusion we want is not conditional.

In a simple case like the one just mentioned we do not (or at least not consciously) invoke a background theory in which to construct the model: to construct a Klein four-group, for instance, it would be enough to point to the symmetries of a rectangle. In more complicated examples, however, mathematicians

[1] Like almost everything else I say in this chapter, this assertion is controversial, at least in the minimal sense that there are philosophers (e.g. Nolan 1997) who deny it. Dialetheists, who hold that contradictions can be true, even go as far as to think that the *actual* world is logically impossible.

do, when constructing counterexamples, invoke the idea that there is a background theory in which the construction is carried out. If they are asked what that background theory is, many of them will say it is ZFC. But in most cases it is very doubtful whether they really mean this. Most of them, most pure mathematicians that is, have only an approximate idea what ZFC is.

Incidentally, many philosophers (even some philosophers of mathematics, who should know better) will be surprised by this last remark, because they imagine that all mathematicians do a course on ZFC as part of their training. But this is not so: many, even many who end up as professional pure mathematicians, do not even do an *informal* course on set theory. (At Cambridge, to take the example I know best, the only set theory course in the undergraduate curriculum that goes beyond a minimal introduction to the language of sets is optional, and most students do not opt for it. There is no requirement to make up this deficiency at the graduate level, and many choose not to do so.)

Of course, the fact that mathematicians do not know what ZFC is does not in itself show that they do not defer to it as a criterion of correctness. (I recently signed a contract in which I agreed to be bound by the law of Scotland when interpreting it, but I have even less idea about what that really amounts to than most mathematicians have about ZFC.) So we should no doubt treat this as no more than a reason for suspicion. Another reason for suspicion is that even those mathematicians who *do* know what ZFC is mostly make no attempt, in constructing counterexamples, to exploit the power or the intricacies of ZFC. The only counter-instances to this that I can think of occur in parts of mathematics that really are quite close to set theory (most notably general topology).

This is an instance of a phenomenon any philosopher of mathematics, at least any who takes the practice of mathematicians at all seriously, has to get used to, namely that what mathematicians *say* is not always a reliable guide to what they are doing: what they mean and what they say they mean are not always the same. The deference towards ZFC which many mathematicians claim is actually a myth. They do not really believe that the criterion of correctness of a putative theorem is whether a formalized surrogate can be proved in ZFC (as is shown, rather trivially, by the fact that they can be persuaded of the truth of Con(ZFC)).

Of course, I do not mean by this that there is some *other* formal theory (such as ZFC+Con(ZFC)) which provides the context for mathematical theorizing of this sort. My point is only that although when constructing counterexamples mathematicians situate themselves in a background context, we should not too readily understand that context to consist in some particular first-order formal theory. Mathematics journals frequently reject papers because the arguments in them do not convince the referees. They do not reject them because the referees doubt that the arguments can be formalized in ZFC.

If I am right, then, that ZFC is not really the background theory for actual mathematics, how did the myth grow up that it is? To understand this, we have to examine how mathematicians work in practice. Most of the time, a mathematician working in a particular branch of mathematics will make use only of methods that are familiar to, and accepted by, other practitioners in that branch. Occasionally, perhaps, a step in a proof might cause difficulties to other people. Then the mathematician will be asked to explain it in more detail, and the explanation will typically be at a lower level. This process has to stop somewhere, certainly. But where it stops is the point at which all parties to the dispute are agreed, and this will always be very far short of a complete formalization of the argument. The myth arose because from the fact that disputes are resolved by going down a level it was wrongly inferred that there was a final, wholly formal level at which all disputes would be resoluble.

In other cases, a mathematician working in one branch of mathematics might use another branch to construct a counterexample. For instance, a counterexample in analytic number theory might be given by means of a construction in algebraic geometry. But even in these cases the mathematicians' real (and conscious) reason for believing the counterexample would be that they trusted algebraic geometry. Any appeal to ZFC once again drops away as irrelevant.

So the background theory within which mathematicians usually operate is inherently informal. The central point we need to observe about it, though, is that we cannot be implicationist about it as well, because that simply postpones the problem: at some point in the process there needs to be something we can assert as *true*, not just conditionally.

In the 1920s it was thought for a while (by Hilbert, most famously) that proof theory might provide a way out. In order to show that something follows from some premises, we prove it. To show that it does not, we might hope to analyse the combinatorial possibilities encapsulated in the rules of proof of the formal system and thereby demonstrate that no string of symbols constitutes a proof in that system of the proposition in question. But this turned out to be a blind alley: in non-trivial cases the proof-theoretic analysis never delivers the required result without making assumptions about transfinite induction which require a substantial background context of their own.

3 *Mother theories*

So let us focus from now on on arithmetic, analysis, and set theory. There is a key difference between these theories and the theories of groups, Hilbert spaces, etc. that we considered earlier. This difference is that we ordinarily conceive of each of arithmetic, analysis, and set theory as having a unique intended model (the natural numbers, the real numbers, the sets, respectively). Of course, we should not pre-judge the philosophical issues simply by assuming that the conception we ordinarily have of these disciplines is correct. Nonetheless, we need to recognize something that this conception we have of arithmetic, analysis, and set theory entails for the philosophical account we offer, namely that any account of a broadly implicationist shape will have a difficulty with detaching the antecedent of the conditional.

When I talk here of accounts with 'a broadly implicationist shape', I intend to include not just implicationism proper but axiomatic formalism, structuralism, and fictionalism. Where they differ is in the form of the antecedent to be detached, not in the need to detach it. The axiomatic formalist says that terms like '7' and '5' gain their meanings from the role they play in the Peano–Dedekind axioms, and those axioms entail that 7+5=12. The modal structuralist says that if there were a natural number structure it would be the case in that structure that 7+5=12. The fictionalist says that according to the story of arithmetic 7+5=12. But in each case they owe us an explanation of how it follows from their account of the matter that seven apples and five oranges make twelve pieces of fruit.

Of course, proponents of these views all do try (with varying degrees of success) to give such an explanation, but they all need to establish the consistency of the axiomatization as a precondition for the success of the explanation. (Typically they need a little more than bare consistency, but we need not go into that at present.) And the difficulty with this is that nothing internal to the theory provides them with the resources to establish this. So they need an external explanation for consistency. But where is that to come from? Let us call this the *postponement problem* and the positions that face it collectively *postponement* views.

The familiar situation, of course, is the one where we model one mathematical theory in another, but, as we observed earlier, we cannot use that now because it only postpones the difficulty once more. What we have to show is that the theory in question is consistent, not merely that it is consistent if some *other* theory is true. When it comes to the mother theory, that is to say, there is no further background in which to do the modelling: it *is* the background.

At this point, incidentally, nothing much hinges on whether the mother theory counts in the traditional taxonomy of these matters as *mathematical*. Thus, for instance, some postponers have justified the consistency of mathematics by appealing to possible worlds in which mathematics is true: for them the theory of modality is the mother theory, and they now owe an exactly analogous debt in respect to it.

The important thing to see is how difficult the postponement problem is for the views that face it. The central difficulty is that the consistency of a substantive mathematical theory is not a trivial matter. This is graphically illustrated by Gödel's incompleteness theorem, which tells us that the consistency of Peano Arithmetic cannot be proved in Peano Arithmetic itself. One suggestive way of thinking of this is that from the perspective of the theory itself the problem of proving its consistency is infinitely hard. Of course, it does not follow that from *every* perspective the problem is hard. And indeed non-postponement views have a simple argument for consistency: the theory is consistent because it is true in its intended model. But for them the complexity is transferred to the issue of what is involved in grasping the intended model. For non-postponement views, on the other hand, it is intrinsically unlikely that the task of establishing the consistency of the theory will be easy, because they are trying to do this from scratch. I have made this point elsewhere (Potter and Smiley 2001: 334–5) in relation to one particular postponement view (neo-Fregean logicism), but it applies equally to all of them.

4 Understanding and truth

I want now to make a distinction which is very familiar to epistemologists, but I think considering its implications will take us a surprising distance with the question we are dealing with here. The distinction is that between understanding and truth, between what is involved in understanding a sentence of mathematics and what is involved in coming to know that it is true.

In the history of philosophy this distinction has not always seemed obvious. Verificationism was the attempt to link understanding to truth by regarding the meaning of a sentence as consisting simply of conditions under which we might come to know it. Philosophers have, I think, come quite widely to see that this is mistaken, but I do not have to address the issue here in its full generality. It is enough for current purposes to notice that in the case of mathematics the view is especially unpromising. Wittgenstein in his middle period, when he was in the grip of verificationism, did indeed try out the idea that the meaning of an arithmetical generalization consists in its proof, but he never made any real progress with the twin problems that this view faces: on the one hand of explaining how we apparently understand arithmetical sentences, such as Goldbach's conjecture, which we currently have not the least idea how to prove; and on the other of what to say about cases where we have two completely different proofs of the same theorem.

So in the case of arithmetic, for example, there are two distinct issues, one of explaining our grasp of the concepts involved, such as addition, the other of explaining our knowledge of the truths we express using those concepts. Notice, though, that the two issues are to some extent interrelated, since which concepts we have is partly dependent on what we know. Of course which concepts we have depends to some extent on our environment, both on what is there and on what we find salient. So of course reflection on our concepts is one route by which we can come to some (fairly limited) knowledge about that environment. If there were not, and had never been, any water, we would not have the concept of water that we have. So someone who reflects on his grasp of the concept of water can thereby come to know that there is water. If there were not, and had never been, any antelopes, we would not have the concept of an antelope. So someone who reflects on his possession of the concept of an antelope can thereby arrive at the realization that there are such things as antelopes. Someone who, not realizing that unicorns are mythical creatures, uses a similar argument to reach the conclusion that there are such things as unicorns is simply making a mistake.

To those involved in the recent debate about the so-called armchair knowledge problem (e.g. Davies 2000; Brown 2003; Beebee 2001) what I have just said will probably seem much too swift. I cannot say much more than that it does not seem too swift to me. The particular cases I have mentioned are of course not strictly relevant to the issue at hand here, since our knowledge that there is such stuff as water is not mathematical. What is relevant is only to note that there are such things as *armchair concepts*, concepts, that is to say, such that reflection on our grasp of them is capable of leading us to non-trivial knowledge. *Water* is, I claim, one such concept, but there are many, many others: Kripke's example of the length of the metre stick in Paris shows that *the metre* is one, for instance.

Notice, though, that although reflection on armchair concepts may lead to knowledge, it is not *new* knowledge. Let us suppose that through reflection I take stock of some things I already know about the concept of water and my possession of it, and thereby come to appreciate that if there had been no water, I would not have had the concept. This is an argumentative route that can lead me to the conclusion that there is water. But of course I already knew that. What it is not, plainly, is a route that could be used to enable someone to *acquire* the knowledge that there is water. If you did not know that there is water, you would not have the concept *water*, so what I would have to do to lead you to that knowledge would be to introduce you to the concept. Even if we are free of any desire to mollify those who make much of supposed problems for self-knowledge arising from externalist views of content, therefore, we ought not to speak as if simply by reflection on the concept of water we can *acquire* the knowledge that there is water. It is one thing for us to possess an armchair concept, quite another for this possession to be an independent route to new knowledge.

5 *Why externalism?*

Armchair knowledge was supposed to be a problem for semantic externalism. As I have made clear, I do not myself think that it is. However, although that removes a possible objection, it does not in itself give us a reason to accept externalism. Why should we think that there are *any* armchair concepts? The internalist can accept that our thoughts are made up of components to which we have reflective access, but will insist that these components are purely mental. What is wrong with that?

I cannot in the space available to me here give enough argument to convince the internalist that he is wrong, but I can at least indicate how I think the issue should be decided. The debate between these two positions, semantic externalism and semantic internalism, is the modern heir of the disagreement between Frege and Russell over the reference of singular terms. Externalism is, to this extent at least, the heir of Russell's atomism. Now Russell's own reason for coming to atomism in the first place may well have been little more than an inchoate sense that the internalist alternative was really a form of idealism. But Russell's pupil, Wittgenstein, gave a slightly better argument for externalism in the *Tractatus*.

If the world had no substance, then whether a proposition had sense would depend on whether another proposition was true.

It would then be impossible to form a picture of the world (true or false). (Wittgenstein 1922; corrected edn. 1933: 2.0211–2.0212)

Wittgenstein's point was that we cannot say anything about the world unless something is simply presupposed. If all our concepts were, in the sense at issue, internal, i.e. if the concepts on their own did not determine whether they referred, then nothing we said would be capable of truth or falsity.

It may seem odd that I am quoting Wittgenstein's argument for substance here. After all, this is the key step in the argument for atomism, and if anything in the *Tractatus* is universally agreed to be wrong, it is surely its adherence to atomism. But this is where the point I mentioned earlier about mistakes is important. What Russell (and, following him Wittgenstein) got wrong was their treatment of mistakes. There have been people who thought 'Vulcan' was the name of a planet. As it turns out, they were wrong. Although it looks like a name, it is actually a description, attempting to pick out a planet where there is in fact no such thing. But the existence of almost anything that I have a name for could, if I look at matters from a sufficiently sceptical perspective, be a matter of doubt. So, Russell thought, almost all the apparent names in my language are really disguised descriptions; and, correspondingly, almost all the concepts in my language are really compound. Russell's attempts to identify the items that lie beyond this, the items whose existence is beyond rational doubt, led him to his now thoroughly discredited theory (or, rather, theories) of sense-data.

But what Russell had failed to recognize was that when I get myself into the state in which I doubt the existence of the woman standing right next to me, I have temporarily re-configured my conceptual scheme. Her name is *now*, while I doubt her existence, a disguised description, but most of the time it is not.

What I have just said is no doubt not enough to dispel all concerns about the problematic nature of mistakes. (If it were, Russell would probably have realized it for himself.) But the problem of mistakes does not have the significance that it is sometimes taken to have. Consider the case in which I have what I wrongly suppose is a thought about Vulcan. (Or, again, suppose that I have what I wrongly suppose is a thought about the woman I mistakenly believe I can see on the other side of the darkened room.) It is in cases such as these that the internalist seems to be on the strongest ground. The externalist is required to say that I am not having a thought at all. But some sort of mental activity is going on: I am adopting the attitude of belief to *something*. Why can I not just call this something a pseudo-thought and be done with it?

The difficulty with this is that it is hard to see what a pseudo-thought is like. Internally, it is indistinguishable from the genuine thought I would have had if there had been a woman on the other side of the room. How can this be? In recent writings on the subject (e.g. Bell 1988) some effort has gone into studying this sort of case in the hope that we can thereby learn something about the structure of thoughts. But I strongly suspect that this effort is misdirected.

To see why, consider another example of a mistake. Suppose I play what I and my opponent believe is a tennis match. Unknown to both of us, however, the court on which we played the game did not quite conform to the specification in the rules of tennis. If, after the mistake comes to light, you ask me what the score was, how should I reply? Strictly speaking, what we were doing was not playing tennis. We were hitting a ball backwards and forwards across a net, no doubt, but that is not enough for it to be tennis. Strictly speaking, the match had no score, because it was not tennis.

It is easy to imagine you becoming impatient with my persistent refusal to tell you the score. (I know I would if I were you.) Perhaps you resort to a counterfactual. What *would* the match score have been, you ask me, if the court *had* been the correct size? But this is not an easy question to answer. If the court had been a different size from the size it actually was, it would not have been the same match. Perhaps a couple of crucial balls, which in fact fell just out, would on the different sized court have just clipped the line, with consequent effects on the score. That, I imagine, is not what you want to know. Nor will it do if you ask me what I thought the score was when the match finished, before I discovered the mistake in the layout of the court. You could ask me, perhaps, what the umpire *said* the score was, but even that is not quite right. What if the umpire forgot to read the score out? What if you reproduce all the rules of tennis apart from the one about the size of the court, and name the game thus specified pseudo-tennis? Having done so, you ask me what, judged according to *these* rules, the score was. But even that is not quite right. The umpire judged according to the rules of tennis. Is it so clear that if he had been judging according to the rules of pseudo-tennis, all his decisions would have been the same?

I do not want to labour the example too much. Its moral, it seems to me, is this. We have elaborate languages devised to deal with *successful* cases of particular kinds of activity. When we encounter a failed case, something that almost lies within the scope of such a language but for some reason falls just short, we are often at a loss how to describe it. There is something genuinely puzzling about this. But what is puzzling about it is nothing particular to tennis. Fairly obviously I could have used any number of other examples, which would have displayed just the same overall shape. One of those examples is the one mentioned earlier of something that just fails to be a thought. If we do not think that a detailed study of the pseudo-tennis case is likely to be profitable in telling us anything illuminating about tennis, why should we imagine that reflecting on pseudo-thought might tell us anything useful about thought?

6 *The route to knowledge*

There is another distinction I want to mention, that between how I have come to know something and an original or primary route by which it might become known. If you ask how I came to know some mathematical truth, the answer will almost always be a philosophically disappointing one, namely that I read it in a book, was told it by a teacher, or saw it on the screen of a calculating machine. And I am not unusual in that: those will be the answers almost all of us give for almost all pieces of mathematical knowledge. The reliability of these routes to knowledge depends, most of the time at least,[2] on empirical facts (that the book is from a reputable publisher, perhaps, or that the calculating machine has been found to be otherwise reliable). The route to mathematical knowledge is almost always, in other words, a posteriori.

Nonetheless it is common to insist that mathematical knowledge is a priori. What is meant by this, of course, is not that I in fact came to it independent of experience, or even that someone at some time came to it independent of experience, but only that someone could have done.

Notice, though, how bound up this notion is with issues of modality. Many mathematical theorems that the mathematical community regards as known have proofs which, if written out completely, would be far too long for any one person to have had a full grasp of them. The most that one might claim for such proofs is that they consist of chains of subproofs, each of which has been, at least temporarily, grasped by some people (in many cases only the author of the paper and the referees). So if we describe these theorems as knowable a priori, we probably do not mean that anyone really could come to know them independent of experience. The modality here is not a practical one, but a question of what an idealized reasoner could come to know.

7 *Benacerraf's problem*

One way of posing the problem of mathematical knowledge that has become standard is due to Paul Benacerraf. So standard has it become, in fact, that it is nowadays a painful cliché for articles on mathematical epistemology to begin by stating 'Benacerraf's problem'. What we want, it is said, is a naturalistic account of the epistemology of mathematics, and any such account will involve a causal connection between the knower and the objects known about. But the surface syntax of mathematical statements makes it seem that the terms in them refer to mathematical objects. Mathematical objects are abstract and therefore cannot participate in causal chains. So we cannot know about mathematical objects.[3]

Thus Benacerraf's problem. But it seems to me to be a thoroughly misleading way of putting the issue, and to encourage thoroughly unhelpful ways of thinking about it. Let me explain why. Note first that it is actually quite hard to make the view precise in such a way as to explain what role the mention of *objects* is playing. The reason for this is that aboutness is much more slippery than people tend to think. What is the law of supply and demand in economics about? Physicalists will have to say it is really about electrons and protons. But which electrons and

[2]For an argument that such routes can *sometimes* be a priori see Burge (1993; 1998).

[3]For more on Benacerraf's problem see the Introduction to this volume.

protons? What about future configurations, or possible configurations? Consider 'Either it's raining or it isn't'. What is this about? According to Wittgenstein's theory in the *Tractatus* it is not about anything. Is that right?

Recalling the earlier distinction between the meaning of a sentence and what is involved in verifying it, note that there will correspondingly be two notions of aboutness: first, what is involved in grasping the proposition; second, what is involved in coming to know it. It is presumably the second of these that is relevant to Benacerraf's problem, and for this it is not obvious that there need be *anything* a proposition is about: different routes to verifying it may involve different objects, and there may be no intersection between them.

I am not denying that there is *a* sense in which $7+5=12$ is about the number seven (among other things). What I am denying is that this sense need be the only one relevant to epistemology. (Compare the sentence about the rain. To know that it is true I don't need to have ever seen rain. What I need to know is only that *rain* is a concept in good standing of a particular sort. But that might be regarded as purely grammatical information.)

Suppose, though, that we could resolve these issues satisfactorily, and it emerged that there is a moderately stable sense of aboutness according to which we can identify at least some of the objects mathematical sentences are about. Even then it would not be clear what role the objects should be expected to play in causation. To make an obvious point, we may dispute whether the relata of causation are facts or events, but what they are not is objects: talk of objects as either causes or effects is at best a loose way of talking and at worst a category mistake.

So to ask how abstract objects can be causes is misleading: what is meant is whether the facts or events that abstract objects are involved in can be causes. But if you believe in abstract entities at all, you surely cannot think that the facts or events that are causes should be expected to have *no* abstract constituents to them. So the problematic case can only be that of facts or events *all* of whose constituents are abstract, i.e. ones with no concrete constituents at all. But remember that what we are after here is a problem for *mathematical* knowledge that is not also a problem for logical knowledge. How does a mathematical fact differ from a logical one? One popular view is that there are no logical objects, only logical concepts. On the other hand, the standard view is that there *are* mathematical objects (natural numbers, real numbers, sets). So presumably this is the allegedly relevant difference: the problematic case, that is to say, must be that of a cause all of whose constituents are abstract and at least one of which is an object. But there is work to do to explain why this case should be thought especially problematic.

Suppose, though, that we grant all this. Why think that there has to be any causal connection between a fact involving an object and the event that constitutes my coming to know this fact (or the fact that constitutes my knowing it)? The *locus classicus* for the claim that there has to be such a connection is W. D. Hart.

Granted just conservation of energy, then, whatever your views on the mind–body problem, you must not deny that when you learn something about an object, there is a change in you. Granted conservation of energy, such a change can be accounted for only by some sort of transmission of energy from, ultimately, your environment to, at least proximately, your brain. And I do not see how what you learned about that object can be *about* that object (rather than some other) unless at least part of the energy that changed your state came from that object. It is all very well to point out that the best and (thus) true explanation of our state changes in learning probably requires the postulation of objects, like numbers, which cannot emit energy, but about which we nevertheless have beliefs. For this still leaves unexplained how our beliefs could be about energetically inert objects. (1977: 125)

But as it stands, this is hopeless. To take just one obvious example, there are quite a few things I know about various objects outside my light cone. I know, for instance, that they (or many of them) obey the laws of physics (at least approximately). Consider tomorrow's sunset. There is no causal chain from this event to my current state, and yet I know various mundane things about it—when it will happen, in which direction from where I am now, perhaps even approximately what it will look like.

The standard response to this is to say that we should not demand a chain *from* the event known about to the current state of the knower, since that rules out knowledge of particular events in the future. It may suffice, it is suggested, for the state of knowledge and the event known about to have a *common* cause. But why should this be of any help? Why should the common cause suffice to make one into knowledge of the other?

8 *Benacerraf's problem generalized*

So much the worse, one might say, for a causal theory of knowledge. But without such a theory, what is left of the thought that there is anything especially problematic about knowledge of abstract objects? At this point the inheritors of the Benacerrafian tradition are apt to generalize the difficulty. It is not, they say, a problem about causal knowledge but merely about natural knowledge.

It is a crime against the intellect to try to mask the problem of naturalizing the epistemology of mathematics with philosophical razzle-dazzle. Superficial worries about the intellectual hygiene of causal theories of knowledge are irrelevant to and misleading from this problem, for the problem is not so much about causality as about the very possibility of natural knowledge about abstract objects. (Hart 1977: 125–6)

Now 'naturalism' is an often-used word in recent philosophy,[4] but what it means is sometimes not as clear as it should be. One thing that is striking about the way it is used, for instance, is that quite often it really seems to mean physicalism. But I think we can put that behind us quite quickly. If *that* is what is meant, why on earth should we believe it? Why, that is to say, imagine that the methods of physics rather than those of any other science provide any sort of guidance as to the sort of epistemology we should adopt? It is true that some parts of physics offer predictions of remarkable precision and reliability, but these are certainly

[4]For a detailed discussion see Weir (2005).

not uniquely distinguishing features of physics: other sciences make precise predictions too. Moreover, even in the case of inexact sciences such as sociology, it would surely be an error to make the slide from our lack of knowledge of exact facts to supposing that there is nothing at all that we know. And in that case the methods by which we arrive at inexact knowledge, for instance in sociology, may well be just as respectable as those we use in physics, even if they turn out to be different.

So let us suppose from now on that it is a broader naturalism we mean. Naturalism has sometimes been offered as a guide in issues connected with ontology: what we should think there is is just what scientists tell us there is. (That is not to say definitively that there *is* nothing else—the pessimistic induction would obviously make that very unwise—but only that scientists give our best current guess as to what there is.) Straightaway, though, this beguilingly simple dictum must be qualified. At the very least it should be revised to say 'scientists speaking in their role as scientists'. But that seems to mean 'scientists speaking within their sphere of expertise', which begs the question of what their sphere of expertise is. The issue that concerns us is precisely whether their sphere of expertise extends to ontology. It is not even as obvious as some naturalists suppose that scientists are the experts on the ontological commitments of their own sciences. Physicists are alleged to be the experts on what physical objects there are, and biologists on what creatures there are. Are theologians the experts on what gods there are?

Whether we extend the argument to theologians seems to depend on whether they are scientists. On the face of it this is absurd. But why? One answer sometimes given is that there are distinctively scientific norms which theologians (or literary critics) do not adhere to, but which natural scientists (when performing their roles as natural scientists) do adhere to. What are they?

Experiment certainly has a role in some sciences. But experiment (at least as it is normally understood) entails interaction. 'If you can spray them then they are real' (Hacking 1983: 23). And that excludes quite a few of what are usually categorized as sciences. Astronomy is an obvious example. One could, of course, take the heroic course of denying, solely for this reason, that astronomy is a science, but that does seem ill-advised. (Apart from anything else, this shows that Hacking's slogan is far from being a *criterion* for reality:[5] we do not think that our inability to manipulate distant stars shows that they are not real.)

So if experiment is not characteristic of the natural sciences, what about observation? This certainly plays a role even in astronomy and in economics. But isn't it now too broad a categorization? Surely observations (or at any rate empirical data) are relevant not just in science but in any activity whatever.

The general point is that it is quite difficult to characterize the natural sciences by finding distinctively *scientific* norms which they all adhere to. When one tries to formulate them, what one comes up with tend merely to be *rational* norms. Not everything that is rational is part of the natural sciences, and what distin-

[5]Hacking himself did not suggest that it was.

guishes the sciences from other forms of rational inquiry is more to do with the subject matter and the company scientists keep than with anything distinctive about the *epistemological* norms they adhere to.

The contrary view seems to me to have been borne of a false contrast according to which what is opposed to science is 'mere'—mere astrology perhaps, or mere theology. We may agree that astrology is in this sense mere: the claims made in astrology are not knowledge. But it is much too big a jump to say that everything that is not science is in astrology's boat.

The point I am making here is not (or at least is not intended to be) *anti*-scientific. Rather is it that I think some naturalists rely on a false contrast. If I do not put science first, and am willing to subject what it says to extra-scientific critique, they seem to think I put myself at imminent risk of succumbing to a belief in creationism (or horoscopes, or crop circles, or whatever other bête noire is supposed to lurk in the shadows just beyond science's reach). But this is naive: what is not science is not thereby irrational.

One might be tempted, though, to generalize Benacerraf's problem still further (and, thereby, give it a much older heritage, reaching back to Plato's *Meno*). Perhaps, one might say, the problem is not about whether the epistemology of abstract objects deserves to be called naturalistic but only about whether there is any sense in which it tracks the objects at all. No belief we possess deserves to be called knowledge, we might say, unless it covaries counterfactually with what is known: if it were not true, I would not know it. And the problem for mathematical knowledge is just that it does not vary: the counterfactual gets no grip because if a mathematical sentence is true we cannot make the required sense of supposing it not to be.

But that will not do for current purposes, simply because the problem has now been generalized to the point that it applies to *any* necessary truth. It no longer holds any terror for the mathematician that it does not also hold for the logician, whereas what we are after here is what *distinguishes* the mathematical from the logical case.

9 *The real problem*

What, then, is the real problem of mathematical knowledge? It will be clear from what I said earlier that I think the postponement problem is a serious one for the views which face it (what I called postponement views); indeed, I think that it is fatal, although I have not said enough here to demonstrate that. The reason why non-postponement views do not face this problem is of course that they have a quick answer to the question why the theories they are dealing with are consistent: they are consistent because they are true about the intended model.

The real problem for non-postponement views is therefore to explain what a conception of a mathematical structure amounts to in such a way as to make it plausible that we might come to know that certain things are true about it. But, as my scepticism concerning Benacerraf's problem should make clear, I do not think that what makes this especially problematic is that we conceive of the structure as abstract.

That is not to say, of course, that knowing about numbers is just like knowing about tables and chairs, or even that it is just like knowing that either it is raining or it isn't. But the point at which the epistemological problem becomes distinctively mathematical is when we invoke the idea of reflection.

This is why I mentioned armchair knowledge. What the examples discussed earlier showed was that there are concepts, armchair concepts I called them, our possession of which entails facts which are neither trivial nor in any ordinary sense about us. Our possession of the concept *water*, for instance, entails that there is water. What I want to claim is that some mathematical concepts, most prominently arithmetical and geometrical concepts, are armchair concepts in this sense.

One might be tempted to say not just 'in this sense' but 'in this way'. That would be too strong, however. That *water* is an armchair concept follows from semantic externalism. And semantic externalism about this concept is forced by consideration of how our concepts would differ if we lived (and had always lived) on Twin Earth (Putnam 1975). That *number* is an armchair concept does not follow from semantic externalism: one cannot be a semantic externalist about the concept *number*, because the counterfactual account of what semantic externalism amounts to gets no grip on it. Any attempt to explain platonism about mathematics by means of a counterfactual anything like the Twin Earth case is lame from the start.

The challenge which mathematics presents to the epistemologist is therefore to explain how *four* is an armchair concept—how, that is to say, it resembles concepts like *water*, but also how it differs. I am not going to answer this challenge here, though not for reasons of space but because I do not have an answer. What I will do is to point out two general ways of investigating the question that suggest themselves, one potentially more fruitful than the other.

The first way, the one I believe is less fruitful, starts from an obvious difference between the concept *water* and the concept *four*, namely that perception of water (not necessarily perception by me, of course) is involved in my acquisition of the former, but perception of the number four is not involved in my acquisition of the latter. This is just a variant of Benacerraf's problem, and I have tried to explain earlier why I do not find that very interesting. But the fact that I do not find an approach interesting is certainly no good reason for others not to pursue it. Why am I so sure that this approach is unlikely to bear fruit? A very short answer would be that Benacerraf's problem, if it is a problem at all, is a problem for the epistemology of *any* sphere of discourse in which the objects referred to are abstract, whereas mathematics has features which seem to throw up quite special difficulties not found in other spheres. For a slightly fuller answer let me contrast this approach to the other one, which I believe is likely to be more fruitful.

This second approach focuses on the tension between two competing intuitions. The first of these is broadly internalist in its drift. To see how it arises, consider a particular case, namely the worrying similarity between Hume's principle[6] and Basic Law V.[7] These two principles share a syntactic form (that of what are known as abstraction principles). The first is consistent and from it we can, if we use second-order logic, derive the whole of arithmetic. From the second we can, again using second-order logic, derive Russell's paradox. Is either of these principles a means of introducing us to an armchair concept? The second plainly is not (dialetheism apart). The first, though, might be. Where, if so, does the difference between the two lie?

Earlier, I stressed that semantic externalism about empirical concepts requires us to have a relaxed attitude to mistakes. Should we adopt a correspondingly relaxed attitude to non-empirical concepts such as *number*? Should we, that is to say, try to argue as follows? There is no need for us to identify any feature internal to the concept introduced by Hume's principle that makes it a successful as a route to numbers. The process by which we come to grasp the armchair concept introduced to us by Hume's principle is indeed *just* the same as the process by which we come to grasp the contradictory concept introduced to us by Basic Law V. What is different lies not in the nature of the concepts but in the structure of the world. There are, as a matter of fact, numbers; there are not, as a matter of fact, classes (at least, not the classes that Basic Law V would commit us to).

This, I hope it is plain, is an externalism of the most extreme kind, too implausible to be countenanced. Even Gödel, who is widely credited with a robust conception of the mathematical realm as quite independent of us, hesitated to accept such a view (see my 2001). Gödel recognized, that is to say, that what makes semantic externalism plausible in the empirical case is our conception of the world as something we can be mistaken about. We can, of course, make mistakes in mathematics (and all too easily), but what a mistake amounts to is quite different. Mistakes in mathematics are, at least in principle, discoverable a priori; the notion of an undetectable mistake in mathematics is surely incoherent.

But suppose now that these considerations lead us instead to give in wholly to an internalist conception of mathematics according to which it owes its truth only to our grasp of the concepts it deploys. If that grasp derives wholly from some principle, whether Hume's or any other, that we can formalize, Gödel's incompleteness theorem tells us that this account is inadequate. For there will be true sentences of our formal language whose truth does not flow solely from the formal principle in question.

The moral here is that the three domains on which the philosophical problem we are discussing centres—the natural numbers, the real numbers, the sets—have, in slightly different ways and to slightly different degrees, a sort of open-endedness. Dummett has in many of his writings used the term 'indefinite ex-

[6]The number of Fs equals the number of Gs if and only if the Fs and the Gs can be put in one-to-one correspondence.

[7]The class of Fs equals the class of Gs if and only if the concepts F and G are materially equivalent.

tensibility' for variants of this open-endedness, but what he has said about it has been closely bound up with his attack on classical logic. Gödel, on the other hand, emphasized its importance in a number of places (e.g. 1944; 1995) without showing any temptation to abandon the law of the excluded middle as a result.

This is surely the point at which the difference between logic and mathematics is at its sharpest. The internalist tendency mentioned earlier is one that could surely be applied to logic just as to mathematics, but the externalist tendency feels distinctly alien there. What gives the epistemology of mathematics its uniquely troubling hue, on the other hand, is the tension between these two competing intuitions, internalist and externalist. It is a tension which thinking about Benacerraf's problem seems to me to be singularly ill-equipped to resolve.[8]

[8]I am very grateful to Alan Millar, Alan Weir, Tim Gowers, the members of a Cambridge discussion group, and an Oxford University Press referee. All provided me with comments on earlier drafts of this chapter which saved me from errors.

Mathematics, Memory, and Mental Arithmetic

W. T. Gowers

1 Introduction

A few months before the conference that led to this volume, I bumped into Alexander Paseau in the street, and he invited me to speak at it. Having always been something of a philosopher manqué, I was extremely flattered; but he quickly brought me down to earth by stressing that the conference was interdisciplinary, and that the discipline he was asking me to represent was mathematics. This raised an interesting problem: what is it to speak *as a mathematician* at a conference on the philosophy of mathematical knowledge? The answer I chose was to subject myself to three constraints, which were designed not to allow me to become too purely philosophical.

This article will follow closely what I said in the lecture, though without necessarily sticking to it exactly. In particular, the same constraints will apply, and they are as follows.

A. *Metaphysics should be avoided.*
I will leave completely alone questions about the existence of numbers or the nature of mathematical truth. Opinions vary about how important such questions are, but it is generally conceded that they are questions for philosophers rather than for mathematicians.

B. *The topics discussed should have a potential impact on mathematics.*
Although very few mathematicians pay attention to traditional problems in the philosophy of mathematics, this does not mean that mathematicians are unphilosophical. This article will contain several examples of questions, the answers to which could have a genuine effect on mathematical practice, and for each one I will defend the thesis that it is philosophically interesting as well.

C. *Illustrative examples should be taken from 'real' mathematics, and not just set theory and very elementary arithmetic.*
Philosophers of mathematics have good reasons for focusing on the foundations of mathematics, which means that the mathematical statements they discuss tend to be very simple. However, most mathematicians are not interested in foundations: they regard foundational problems as having been solved several decades ago, at least to the extent that one can do 'ordinary' mathematics in complete confidence without having to worry about them. Therefore, to engage the at-

tention of mathematicians one should discuss more complicated statements with actual mathematical interest. (Of course, since this chapter is aimed principally at non-mathematicians, I will keep the examples fairly straightforward, just as, when I discuss philosophical or psychological matters, some of what I say will be very well known to philosophers or psychologists. But I do want to discuss at least some mathematical statements that do not have instantly obvious proofs.)

The above constraints rule out many topics, but they still leave me with a huge area to explore, so huge in fact that I shall have to narrow my attention artificially. But before I do that, let me give some examples of the kinds of questions that should interest both mathematicians and philosophers (and do interest increasing numbers of them).

2 *The problem of induction in mathematics*

By 'induction', I do not mean mathematical induction, but the induction that is discussed by philosophers: under what circumstances are confirming instances of a mathematical statement considered to be good evidence for that statement?

For example, probably all serious mathematicians believe that, in the long run, each of the digits 0 to 9 occurs in the decimal expansion of π about 10 per cent of the time. The grounds for this belief are based on experimental evidence in two ways. First, experience suggests that no naturally occurring irrational number, such as e, π or $\sqrt{2}$, has any *reason* to be biased towards one digit rather than another. Nobody has found a proof, or even an informal argument, that any of these numbers is biased, and nobody expects to. This is significant because there seems to be a very general, and not wholly precise, principle operating in mathematics, which says that no truly striking statement is ever true unless it is true for a good reason. A second and more obvious (but not necessarily more convincing) piece of evidence is that millions of digits of π itself have been calculated, and not only have they shown no bias, but they have also passed many other statistical tests for randomness.

As far as formal proofs go, almost nothing is known about the decimal expansion of π. Nobody has ruled out the possibility that every digit after the 10^{100}th is either a 7 or an 8. And yet it is clearly ludicrous to suppose that that might be the case.

In the other direction, consider the decimal expansion of e, which begins 2.718281828.... It is quite striking that a pattern of four digits should repeat itself so soon—if you chose a random sequence of digits then the chances of such a pattern appearing would be one in several thousand—and yet this phenomenon is universally regarded as an amusing coincidence, a fact that does not demand an explanation. Why?

Induction of the scientific kind plays no part in the formal validation of mathematical statements. However, it is vitally important to research in mathematics.

For example, suppose that one is trying to prove a theorem that states that every natural number n has a certain property. If the problem is a hard one, then it may well be necessary to identify an intermediate statement, prove that, and then deduce the theorem from it. But even the intermediate statement will often be hard to prove, so before trying to do so, it is a good idea to try to find convincing but informal evidence that it is true. Induction is one source of such evidence, so the more one understands it, the better one will be at devising good research strategies.

The problem of induction in mathematics is also fascinating as a purely philosophical question. It closely resembles Goodman's 'second riddle of induction,'[1] but it takes a particularly pure form in this mathematical context. One is not distracted by worries about the nature of causality, or of physical laws, or of the reliability of observations, and yet the basic problem—when do we think that confirming instances of a general statement provide genuine supporting evidence for that statement?—remains. Indeed, I would maintain that for this reason the problem is *more* interesting in a mathematical context. Without the distractions from the 'real world', there seems to be more chance of making progress, and such progress ought to shed light on the more traditional problems of induction.

However, I shall say no more about this because it is the subject of Alan Baker's contribution to this volume.

3 *The seemingly inappropriate use of modal language by mathematicians*

In 1993, Andrew Wiles announced that he had proved Fermat's Last Theorem. It turned out that his proof was not correct, but with the help of Richard Taylor he was able to make it correct a year later, and thereby solve the most famous problem in mathematics. At one point, between the first announcement and the eventual proof, a rumour spread around the internet that Noam Elkies had found a counterexample to the theorem. Had the rumour been correct, it would have been quite astonishing, but the rumour had some plausibility because Elkies had already astonished mathematicians by disproving a superficially similar and widely believed conjecture of Euler.

There is nothing particularly controversial about the last sentence of the previous paragraph, but for a philosopher it presents difficulties. What exactly is the status of the counterfactual statement I made when I claimed that it would have been astonishing if Elkies had found a counterexample? After all, now that the

[1]The first problem of induction is to explain why it is rational to believe, as a result of past experience, a statement such as 'If I cross that bridge on a normal day then it will not collapse.' In *Fact, Fiction, and Forecast* Goodman famously observed that for many statements such a belief is not rational. For example, if 'grue' means 'green if first observed before the year 2020, and blue if first observed in 2020 or later', then the fact that grass has generally been grue up to now is not good evidence that it will generally be grue in 2025. Of course, this statement is rather artificial. The second riddle of induction is to explain what makes some statements more 'natural' than others.

theorem has been proved, we know that there *is* no counterexample, which implies that there is no counterexample in any possible world. This makes it look as though the claim is true for the trivial reason that any counterfactual conditional with a necessarily false antecedent is true. But that cannot be a correct analysis. Why do we readily accept that if Elkies had disproved the conjecture then it would have been astonishing, but we do not accept that the moon would have turned blood red?

It is tempting to try to modify the idea of possible worlds by using a looser notion of possibility. Sometimes it is rational to find a statement plausible even if it is logically contradictory, at least if the contradiction is very hard to find. So perhaps one could develop a theory of worlds that are plausible given the current state of knowledge. However, this does not look like an easy task, and it is not clear that it could be done with anything like the neatness one would normally want from such a theory.

Another temptation is to say that we cannot genuinely conceive of a situation where Elkies disproves Fermat's last theorem, simply because the situation is logically contradictory and therefore inconceivable. What we are doing is imagining something different, such as how we would feel if we believed (wrongly, but for good reasons) that Elkies had found a counterexample. But that is not what it feels like when one makes a counterfactual statement in mathematics. If I entertain the possibility Elkies might have found a counterexample, then I want to say that I am talking about *that theorem*, rather than some general scenario, and the possibility that *it turned out to be false*, rather than that I mistakenly believed it to be false.[2]

Counterfactual statements are not the only ones that are puzzling; there are many varieties of informal statements that appear to suggest that the truth or falsity of some mathematical statement is not yet determined. For example, probabilistic language is all-pervasive in discussions between mathematicians. I have already used it in this paper, when I suggested that the occurrence of the pattern 18281828 at the very beginning of the decimal expansion of e was thought of as a 'coincidence'. How can it be a coincidence, when the pattern *necessarily* occurs? Do we implicitly have in mind some probability distribution–e.g. of the expansions to different bases of various important irrational numbers? Similarly, some conjectures are regarded as 'probably true', while others 'could go either way'. These judgements are not pure whimsy: one can argue about them, change one's mind in the light of further evidence, and so on.

It seems likely that Bayesian analysis will be helpful, at least for the purposes of explaining why it is rational to make informal probabilistic judgements. However, even if that general principle can be established, it leaves open the large and worthwhile project of examining how such judgements are, and ought to be, made in practice.

[2]The italics in this last sentence play a similar rhetorical role to Kripke's italics in *Naming and Necessity* when he discusses rigid designators. For example, he urges his readers to accept that by 'Hesperus' he means *that planet*, and not something more complicated to do with how it is presented to us.

4 Explicating informal mathematical terminology

Plausibility is by no means the only informal quality that mathematicians will wish to judge when they are assessing a piece of mathematics. There are many words that do not have precise meanings but which play an essential role in mathematical research. A few of them are 'amusing', 'beautiful', 'because', 'comprehensible', 'deep', 'elegant', 'explanation', 'important', 'idea', 'ingenious' (used of a piece of mathematics rather than of its inventor), 'interesting', 'natural', 'obvious', 'technical', and 'trivial'. Several of these are used for evaluating and describing proofs. From a strictly logical point of view (which is the point of view taken by many traditional philosophers of mathematics) the quality of a proof is not important: if it is correct, then it establishes the theorem, and otherwise it does not. However, when one looks at the actual practice of mathematics, it becomes abundantly clear that proofs are far more than mere certificates of truth. And yet, even quite simple-sounding questions about them can be hard to answer. For example, one often says that two proofs are 'essentially the same'. What does this mean? Sometimes it is obvious that two arguments are not really different proofs at all, but simply different presentations of the same proof. But sometimes the essential sameness is far from obvious, and becomes clear only after a lot of hard thought. To give general criteria for this is a fascinating problem, and as far as I know it has not yet been done in a satisfactory way.

Why should imprecise language be useful to mathematicians? Surely precision is one of the hallmarks of mathematics, a feature of which mathematicians are extremely proud. The answer is that, although the statements and proofs that we deal with are (or can be) expressed precisely, actually finding these statements and proofs is a much less precise process, one that involves making value judgements. And we do not have a formal language for expressing them. The reason such judgements are so useful is that, as centuries of experience have shown, proofs of natural, elegant, interesting statements almost always involve other natural, elegant, interesting statements. Therefore, if one is searching for a proof and is forced to make guesses about how it might look, it is a very good strategy to make guesses that are natural, elegant and interesting, if one possibly can. And if one wants to hone one's strategy, then it will be a good idea to try to understand as well as possible what it means for a guess to be natural, elegant, or interesting. Of course, we like natural, elegant, interesting statements for their own sake too, but it is remarkable how important a well-developed aesthetic sensibility can be, for purely pragmatic reasons, in mathematical research. (Another important feature of such statements is that they are more memorable, a quality that will be discussed later in this chapter.)

One way that one might imagine trying to understand better our evaluations of mathematical statements is by a process of *explication*: that is, by coming up with meanings for words like 'natural' that are sufficiently similar to the usual meanings to justify the use of the same word, but sufficiently precise to be amenable to analysis. (The natural sciences are full of examples of explication. For example, the meaning of the word 'hot' to a physicist does not correspond exactly with one's subjective experience of heat, but the two are very closely related, and the physicist's word is much more precise, in a way that makes it possible to study heat scientifically.)

Explication is one of the major activities of analytic philosophy. Most philosophers are not content with the ordinary day-to-day meanings of words such as 'true', 'believe', 'intend', 'decide', 'know', 'free', or 'able' (the list could be continued almost indefinitely). Rather, they try to find explications of these words—though they would more commonly use the word 'theories'. Therefore, the project of explicating the imprecise language used by mathematicians is one that should appeal not just to mathematicians, but to anybody with the instincts of an analytic philosopher.

5 *Memory*

It will be clear that the problems I have been discussing have something in common. In one way or another they all concern what one might call the linguistic superstructure of mathematics, the high-level language that is far more typical of a conversation between mathematicians than the low-level formalities of a rigorous proof. Behind each question lies the conviction that this high-level language is more precise than it seems, or at least that the work that it is doing can be understood more precisely.

Although I think that such questions have had far less attention than they deserve, I am certainly not the first person to be interested in them. Polya stands out as a mathematician who has made a serious effort to understand the processes of mathematical discovery, including the heuristic arguments that guide people towards rigorous proofs. Imre Lakatos's famous *Proofs and Refutations* reminds us how complicated these processes can be, and how much less tidy the discovery of some mathematical theorems is than the presentation of the finished products in a typical mathematics course. More recently, David Corfield has written a widely discussed book entitled *Towards a Philosophy of Real Mathematics*, in which he makes it clear that he too regards this class of problems as important. I therefore want to do more than merely add my voice to the call for them to be investigated more: I would like to do a bit of actual investigating.

It is at this point that I shall narrow my focus, concentrating on the third class of questions above—the explication of the language used to evaluate proofs.

The difficulty one faces when trying to be precise about words like 'interesting' and 'beautiful' is that they seem very subjective. How can there be a precise meaning to the word 'interesting' when mathematicians disagree, sometimes profoundly, about what constitutes an interesting piece of mathematics?

This difficulty may seem pretty serious, but there is a way round it: one can search for explications that are precise, but also *relative*. Indeed, this is clearly the right thing to do, since it is an experimental fact, and not a surprising one, that the way people perceive mathematical beauty, difficulty or interest depends very much on what mathematics they know. So, while it is probably wrong to look for a useful theory of what makes a mathematical statement interesting *in isolation*, there is no obvious obstacle to developing a theory of what makes a mathematical statement an interesting addition to an existing body of knowledge.

At this point one might raise the possibility that two mathematicians could be experts in the same field, and have a very similar background, but differ in their *tastes*, so that they still disagreed about what was interesting, even in their own speciality. Indeed, it is very likely that this happens, though it would be hard to test it because one would never be quite sure that some small difference in background was not what accounted for the disagreement. However, even if such mathematician-pairs do exist, it could be that a sufficiently convincing explication of the word 'interesting' would influence their way of speaking in the future. People might start having to say things like, 'Well, I see that this statement is interesting in the technical sense, but there's some funny side of me that cannot help being bored by it.' If the technical sense of 'interesting' overlapped enough with its usual sense to gain wide acceptance, then this would be an apology rather than a way of saying that the statement was, despite appearances, boring.

Another difficulty that applies to some words is that, as Wittgenstein forcefully pointed out (see, for example, his discussion of games in *Philosophical Investigations*, beginning at §65), they are used in many different ways. For example, there is no single property that all beautiful objects, beautiful landscapes, beautiful works of art, and beautiful people have in common (unless one wishes to define an artificial property such as 'belongs to the set of all objects, natural phenomena, works of art or people that are commonly perceived to be beautiful'). Probably the same is true even if one restricts one's attention to mathematical proofs. Different people have very different reasons for finding a proof beautiful.

This difficulty is more serious; perhaps most of the informal evaluative words used by mathematicians have meanings that are too diffuse to be explicated satisfactorily. But even that would not rule out a general project of explication. If we want a good theory of informal mathematical evaluation, we are not forced to take all the commonly used words and explicate them one by one. We can if we prefer devise a new vocabulary that does the same job but more precisely. If we can use this new vocabulary to give explications of the old words, then so much the better, but if the new vocabulary works well then we may find that our urge to do so is diminished.

The main suggestion of this paper is that one can get a long way towards such a theory by thinking about *memory*. All other things being equal, a memorable proof is greatly preferable to an unmemorable one. But what is it that makes some mathematical statements and proofs much easier to remember than others?

This question has several good features. First, it is sufficiently precise that one can investigate it and hope to obtain clear results: either you remember how to prove a theorem or you don't, and if you don't then somebody who does will often be able to tell you exactly what part or parts of the proof you have forgotten.

Second, such investigation will be interdisciplinary: psychologists could devise tests to see whether certain features of proofs make them memorable; mathematicians and philosophers could make informed guesses about what such features might be, or else search for features that are interesting in themselves and positively correlated with memorability. Third, whether one finds a statement or proof easy to remember depends on one's mathematical background. Thus, memorability is a relative notion, which makes it more likely to be appropriate for explicating other relative notions. Finally, memorability does seem to be intimately related to other desirable properties of proofs, such as elegance or explanatory power. I shall try to justify this assertion later.

Before thinking about what makes some proofs easy to remember, one should first have some idea of how mathematicians set about memorizing them. In a moment I shall discuss a concrete example, but first let me make the general comment that there is a very close connection between memorizing a proof and understanding it. It is obviously easier to remember a proof if one understands the argument (even if it is hard to say precisely what 'understands the argument' means); it is less obvious, but true, that one's understanding of an argument is greatly advanced if one commits the proof to memory.

There is plenty of evidence for this last claim. Here are two illustrations from my own experience. In Cambridge, we have a very tough one-year mathematics course called Part III. Graduate-level lectures are given in a wide range of subjects, from which a candidate typically chooses six for examination. Most papers take the form of 'bookwork': that is, candidates are asked to reproduce from memory, and at great speed, various theorems, definitions and proofs that have been explained to them in lectures.

Some people complain that the Part III examination is 'merely a test of memory', the suggestion being that it does not test what really matters, namely the ability to do successful research in mathematics. And yet there seems to be a surprisingly strong correlation between how good people are at this memory exercise and how well they do when they go on to do research.

It is not hard to see why this might be. The quantity of material that one is expected to memorize for Part III is so great that there is no hope of remembering it all unless one can find a good way of systematizing it. Instead of remembering the details of a proof, it is much more efficient to remember a few important ideas and develop the technical skill to convert them quickly into a formal proof. And it is better still if the ideas themselves are not so much memorized as *understood*, so that one feels that they arise naturally.

A second piece of evidence is that mathematicians who give lectures without using notes report that this practice improves their understanding of what they are lecturing. Of course, even with notes one cannot give a well-organized lecture without making an effort to understand the material. But if one lectures without notes, the pressure to organize them well is greater, since it becomes important to minimize the amount of straight memorization involved. (It would be interesting to see whether mathematicians with bad memories give clearer lectures than mathematicians with good memories. This could probably be tested in a properly scientific way.)

The fact that memory and understanding are closely linked provides some encouragement for the idea that a study of memory could lie at the heart of an explication of the looser kind described earlier. It is not easy to say precisely what it means to understand a proof (as opposed, say, to being able to follow it line-by-line and see that every step is valid), but easier to say what it means to remember one. Although understanding a proof is not the same as being able to remember it easily, it may be that if we have a good theory of what makes a proof memorable, then this will shed enough light on what it is to understand it that the difference between the two will be relatively unimportant.

6 *Direct memory, competence, and generated memory*

It will make the later discussion clearer if I distinguish between various different aspects of memory. The obvious process that one most readily associates with the word 'memorize' is a sort of direct storage of data. We don't quite know why the data stays there, but if it is of the right kind, and if we are exposed to it enough, then it does. A very different sort of memory is the hardwiring of *competence* at various tasks, such as driving a car: we remember *how* rather than remembering *that*. With practice, we can internalize habits of thought that enable us to perform complicated tasks without planning our actions in advance. This includes not just physical tasks but also intellectual ones, such as solving the equation $2x + 3 = 7$ or having a leisurely philosophical conversation.

One of these tasks is reproducing facts, ideas or fragments of speech, and this leads to what I shall call 'generated' memory. For example, a fact that I happen to be able to recall is that Francis Bacon painted a picture called 'Three studies for figures at the base of a crucifixion.' I did not make a special effort to remember this. Rather, there was a time in my early twenties when his paintings interested me a lot, and this piece of information stuck in my mind. In order to reproduce it for the purposes of writing this paragraph, the first thing I did (not particularly consciously) was to visualize the painting. It is a triptych with a strange animal-like figure in each of the three parts, so the beginning of the title was easy to remember. I also remembered something of the title's paradoxical nature—the figures in question look much more tortured than the onlookers in a typical crucifixion painting—and that brought the rest of the title into the forefront of my mind.[3] As for the artist, nowadays I remember his name directly, but at first I thought of him as the one who shares a name with a famous philosopher. (But is it true that I remember him directly now? There is still a faint echo of the philosopher when I think of the painter.) So what my mind contains seems to

[3]Honesty compels me to admit that this is not completely true: it actually brought to mind the incorrect title, 'Three figures at the base of a crucifixion.' My attention was drawn to the error by Mary Leng.

be a very efficient kind of storage: not of the information itself but of a slightly complicated way to reproduce it.

Nearly all feats of memory are a mixture of both direct and generated elements, but they come in differing proportions. For example, to learn six telephone numbers it feels as though there is no alternative but to store them directly, whereas to learn how to count to a million one directly stores the first few numbers and generates the rest by means of arithmetical rules combined with rules such as these.

1. If $3 \leq n \leq 9$, then one obtains the word for $10n$ from the word for $n+10$ by replacing 'teen' with 'ty', except that the word for 40 is spelt 'forty' rather than 'fourty'.
2. If $2 \leq n \leq 9$ and $1 \leq m \leq 9$ then the word for $10n + m$ is the word for $10n$ followed by a hyphen followed by the word for m.
3. If $1 \leq n \leq 9$ then the expression for $100n$ is the word for n followed by the word 'hundred'.
4. If $1 \leq n \leq 999$ then the expression for $1000n$ is the expression for n followed by the word 'thousand'.
5. If $n = 1$ then one can say 'a' instead of 'one' in rules 3 and 4.
6. If $1 \leq n \leq 9$ and $1 \leq m \leq 99$ then the expression for $100n + m$ is the expression for $100n$ followed by 'and' followed by the word for m.
7. If $1 \leq n \leq 999$ and $1 \leq m \leq 999$ then the expression for $1000n + m$ is the expression for $1000n$ followed by the expression for m.
8. The next number after 999999 is called 'one million' or 'a million'.

However, nothing is quite as pure as it seems. In practice, when learning a telephone number one is helped by a certain amount of background knowledge. For example, I know that the dialling code for Cambridge is 01223, that it is followed by a six-digit number, that for historical reasons the first of these digits is often a 3, that if it is a university number then it is likely to begin 33, that if it doesn't then there are other probable sequences, and so on. Even when learning a random sequence of 11 digits, one may well try to spot little patterns that serve as mnemonics and make the storage not quite direct.

As for counting to a million, it is interesting to think about how one learns the rules 1–8. Are these directly stored? Clearly not. To begin with, one doesn't tend to make them explicit—I had to think a bit before I wrote them down. Secondly, one can make economies by relating some of these rules to others. For example, one could have in mind the general principle that there are aural clues—'thousand', 'hundred and', '-ty' or silence—that indicate which power of ten one is talking about. Then the basic rule is that in order to say a number like abc one says the names of the digits in sequence, following them by the appropriate noise. Of course, this rule is not quite correct so one has to learn exceptions, which themselves may be given by quite general principles—that 0 is typically passed over in silence, that you do not say 'and' if you have nothing to say after it, that teens behave differently, and so on.

Somewhere in between the two extremes comes the learning of a poem or a piece of music. Works of art typically follow many conventions with which we are already familiar, and have internal structural features (such as repeats with small and semi-systematic variations) that make them much easier to remember than arbitrary arrangements of words, notes or images. In the case of poetry and music they are also arranged linearly, which has an interesting effect on how we memorize them. It may be quite hard to remember how the middle of a tune goes unless one first hears the beginning, or plays it in one's mind. If one hears a piece after not having heard it for a long time, then one will often start by remembering just the very beginning, but then find that as one hears each part, it stimulates the memory of the next part so that one can anticipate it just before one hears it. There is clearly something very fundamental about the process of arranging items of memory in a line, with each one linked to the next. A familiar example that illustrates this well is the fact that it is much more difficult to say the alphabet backwards than it is to say it forwards. This example is an interesting one because it shows that even when what we want to remember is an unstructured set, we often remember it by imposing an arbitrary structure on it—as though our brains have a distinct preference for generated memory over direct memory. This is particularly true once we reach adulthood and have filled our minds with arbitrary facts, most notably the vocabulary of our native languages.

7 Two proofs and how one remembers them

Where does memory of proofs fit into this general discussion? There doesn't seem to be a simple answer: some proofs need quite a lot of direct memorization, while others generate themselves, at least in the head of a sufficiently experienced mathematician. But this is encouraging, since it suggests that if we think about what it is that makes memorable proofs memorable, then we may find precise properties that some proofs have and others lack. If we are lucky, these will relate to the informal evaluative terms that mathematicians already use about proofs. Even if they don't, the properties we find may still be useful. For instance, if mathematicians come to understand better what makes proofs memorable, then they may be more inclined to write out memorable proofs, to the great benefit of mathematics.

Rather than looking at this question in the abstract, let us examine a couple of specific examples of proofs and see how a typical mathematician might memorize them. I have chosen the standard proof that every number can be written as a product of prime numbers, and a somewhat less standard proof that the square root of 2 is irrational (but one that is instructive for my purposes here).

The first of these results forms part of the fundamental theorem of arithmetic, which is the statement that every positive integer can be written *in precisely one way* as a product of primes. To put this more formally, if n is a positive integer, then there is exactly one sequence of primes $p_1, \ldots, p_k$ with $p_1 < \ldots < p_k$, and exactly one sequence of positive integers $r_1, \ldots, r_k$, such that

$$n = p_1^{r_1} p_2^{r_2} \ldots p_k^{r_k} .$$

To illustrate: the number 61335 equals $3^2.5.29.47$, and there is no other way of writing 61335 as a product of increasing primes. (To make the theorem true when $n = 1$, one adopts the convention that the 'empty product', where there are *no* primes multiplied together, equals 1. It is easy to see that this is sensible: for example, 3×5 multiplied by no further primes should give us 15, and with this convention it does.)

Notice that the theorem is really a combination of two statements: that it is possible to write n as a product of primes, and that this can be done in only one way. It is the uniqueness of the decomposition that makes the theorem interesting, but to keep things simple I am concentrating just on its existence, which is much easier to prove.

Proof of the existence of a prime factorization. If there is a positive integer that cannot be written as a product of primes, then there must be a smallest such integer. Let this smallest integer be n. Then n is not a prime number (or else we would count it as the 'product' consisting of the single prime n). It follows that $n = ab$ for two positive integers a and b, both smaller than n. We defined n to be the smallest positive integer that could *not* be written as a product of primes, so a and b *can* be written as products of primes. But in that case we can combine those two products and we find that n is also a product of primes. This is a contradiction, so the result is established. □

How is this proof remembered? When I wrote it, I did not have to look it up in a textbook; not even in a virtual textbook that I had stored in my brain. Rather, I knew that I would have no trouble generating it, and this is why. First of all, a very standard move, when one is proving a theorem of the form,

> *For every natural number n, $P(n)$ holds.*

is to begin the proof with,

> *If the theorem is false, then there exists a minimal n such that $P(n)$ does not hold.*

If we have written that line, then clearly we want to look at that minimal n. In our case, $P(n)$ is the statement that n can be written as a product of primes, so what we know about the minimal n is that it cannot be written as a product of primes.

It isn't obvious how to use this property, so let's suppose for now that we have directly stored the next simple line:

> *It follows that n is not prime.*

The rest of the proof now writes itself. First, we say what it means for n not to be prime:

Therefore $n = ab$ *for two positive integers* $a, b < n$.

Next, we ask whether there is anything we can say about a and b. (We don't have to remember to do this: it would be pointless to introduce a and b unless they had some properties we could use.) All we know about them is that they are less than n, so we search back to see whether anything can be deduced from that. And when we see the phrase 'minimal n' we know that something can:

Since n *was minimal,* a *and* b *can be written as products of primes.*

It is then easy to deduce that n can be written as a product of primes. Thus, the most that an experienced mathematician needs to remember of the above proof is that it is a good idea to go from the general statement

n *cannot be written as a product of primes.*

to the more specific one

n *is not a prime.*

For many mathematicians, even this step will not need to be stored directly. Here are two ways it might be generated. The first is to bear in mind that we are trying to deduce that n is a product of primes from the fact that all smaller numbers are products of primes. How could smaller numbers be useful? We would need to take the primes that produce the smaller numbers and use them to produce n. What operations can we perform on products of primes to produce further products of primes? The most obvious one is multiplication. So it will help if n is a product of smaller numbers, which it will be if it is not prime. Once we realize this, then we have generated the line that wasn't quite as automatic as the others.

Alternatively, one might remember that the proof given above is a slightly disguised way of describing the procedure that one is taught at school for splitting a number n up into its prime factors. For example, to factorize the number 61335, one notes first that it is not divisible by 2, but it is divisible by 3: $61335 = 3 \times 20445$. Next, we see that $20445 = 3 \times 6815$. 6815 is not divisible by 3 but it is divisible by 5: $6815 = 5 \times 1363$. Then after a bit of trial and error one finds that $1363 = 29 \times 47$. To describe the process economically, one could say that you start with a number n, find a prime factor p and then repeat the process with the smaller number n/p. Eventually, the number you are looking at will itself be prime and then you are done. This process is more or less what the proof above does, except that instead of looking for prime factors of n it looks for *any* factors of n, which in turn it tries to factorize further, and so on. If one has this insight in the back of one's mind, then once again it is natural to try to write n as a product ab.

What morals can we draw from this example? It is typical of many of the simpler mathematical proofs, in that once one has enough experience there seems to be nothing whatever to memorize: it would be safe to give the proof in a lecture without using notes or doing any preparation. Thus, it demonstrates that some proofs are right at the 'generated' end of the spectrum. In order to remember them, one does not 'commit them to memory', but rather one turns one's brain into the kind of brain that is capable of thinking of them on demand.

How does one do this? Some clues can be found in the discussion above. For instance, I mentioned the general principle that, when trying to prove a statement for every natural number n, it is often good to focus on a putative minimal counterexample. Heuristic principles of this kind, which one learns from experience or from being well taught, are very helpful.

A very general principle, which might seem too obvious to be worth mentioning, is that one should always be consciously aware of what it is one is trying to prove, and of what one already knows. A consequence of this, which is less obvious (not in the logical sense, but in the sense that mathematics students are often not as aware of it as they should be), is that if your hypotheses are weak, then this may very well make it *easier* to find a proof, because you have fewer options, and therefore more of your moves are forced.

(A very nice example of this phenomenon is the following problem. The solution can be found at the end of this article, but if you look it up rather than working it out for yourself, then your life will be the poorer for it. Call a real number *repetitive* if for every positive integer n you can find a block of n digits that occurs at least twice in its decimal representation. The problem is to prove that if x is any repetitive number, then so is x^2.)

In the proof given above, this principle helps to generate the line

Since n was minimal, a and b can be written as products of primes.

There was no need to think in advance about whether this was going to be a useful thing to say: we knew *nothing else* about a and b, so it was almost certain that either this line was going to help or the theorem was untrue. (For some proofs, there is another possibility: that the hypotheses one has available are much stronger than one needs to prove the conclusion. However, there was not much scope for weakening them here.)

Let us turn now to our second proof.

Proof that the square root of 2 is irrational. Suppose that $\sqrt{2}$ is rational. Then we can find positive integers p and q such that $\sqrt{2} = p/q$. Let us suppose that we have done so, and that we have written the fraction p/q in its lowest terms. Now $\sqrt{2} = \frac{2-\sqrt{2}}{\sqrt{2}-1}$ (since each term on the top of the fraction is $\sqrt{2}$ times the corresponding term on the bottom). Substituting in p/q for $\sqrt{2}$ we deduce that

$$\frac{p}{q} = \frac{2-p/q}{p/q-1} = \frac{2q-p}{p-q}.$$

Since $p/q = \sqrt{2}$, which lies between 1 and 2, we find that $q < p < 2q$. It follows that $2q - p < p$ and $p - q < q$. Therefore, we have found a fraction equal to p/q but with smaller numerator and denominator. This contradicts our assumption that we had written p/q in its lowest terms. □

From the point of view of memorization, this proof is noticeably different from the previous one. The basic idea, that we should start with a fraction p/q that equals $\sqrt{2}$ and build from it a smaller such fraction, is another instance of the look-for-a-minimal-counterexample principle, and as such is a generated memory rather than a stored one. However, the particular choice, $(2q - p)/(p - q)$, seems to spring from nowhere—or rather, to be generated by the observation that $\sqrt{2} = \frac{2-\sqrt{2}}{\sqrt{2}-1}$, which itself springs from nowhere. For many people, the most obvious way to remember this argument would be to store directly the fact that this particular observation turns out to be useful.

Those with more mathematical experience do not store the proof in this way: it turns out to be connected with the continued-fraction expansion of $\sqrt{2}$. But let us ignore this, regard the step as slightly mysterious, and ask ourselves how we would like to describe the proof. In particular, let us consider it in relation to three of the words from the list of informal terms given earlier: 'explanation', 'natural' and 'idea'.

Does the proof constitute an *explanation* of the irrationality of $\sqrt{2}$? Obviously, one cannot give a definitive answer to this question without a precise idea of what constitutes an explanation, and that is something we do not yet have, but we can at least follow our pre-theoretic instincts and see where they lead. Speaking for myself (but I hope that my instincts are typical), I want to say that it gives only a *partial* answer to the question, 'Why is $\sqrt{2}$ irrational?' In the back of my mind is a conversation that goes something like this.

Q. Why is the square root of 2 irrational?

A. Because if you were able to write it as a fraction, then you would be able to find a smaller fraction that did the job.

Q. Why does that imply that the square root of 2 is irrational?

A. Because you could then repeat the process, obtaining an infinite sequence of smaller and smaller fractions—which is obviously impossible.

Q. Oh yes. But you haven't told me why you would be able to construct this sequence.

A. Look—here's a method that works.

Q. Yes, but *why* is there a method that works?

Q is satisfied with the general scheme of proof—the method of infinite descent, as it is known—and to that extent A has provided an explanation. But something is still lacking: there is a lingering feeling that $\sqrt{2}$ 'might have turned out to be rational', that it was a 'fluke' that the formula $\sqrt{2} = (2 - \sqrt{2})/(\sqrt{2} - 1)$ existed and was useful. Notice that the point where we feel slightly unsatisfied with the proof, if we are looking for an explanation rather than a mere certificate of correctness, is exactly the point where we have to use direct storage to memorize it. This supports the idea that the notion of understanding a piece of mathematics is closely related to how we memorize it. Much the same could be said about the word 'natural'. The proof is natural, apart from the line that we are forced to memorize directly, which feels artificial. There seems to be a close connection between the naturalness of a proof and the ease with which we can remember it in an efficient, generated way.

One final question I would like to ask about this proof is whether it 'contains an idea'. Once again, I am drawn to the artificial step. If a step of some argument more or less generates itself, then it cannot really be what mathematicians normally mean by an idea, so if this proof has an idea in it, then the artificial step is where we should look for it. And it is quite tempting to say that that step *is* the idea. In other words, the idea is simply that one should consider a certain identity. But the artificiality of the step creates a different temptation as well, because ideas do not spring from nowhere. One feels, rightly in this case, that the 'real idea' lurks behind its manifestation in this presentation of the argument, concealed from view.

This creates a puzzle. A step that is too easily generated does not count as an idea, but if a step appears to spring from nowhere, then we are still reluctant to call it an idea. So what is an idea? Is it something like a step that is only half understood? This would make 'idea' a highly relative notion, rather than an intrinsic property of proofs. But perhaps that is correct: certainly, all mathematicians have had the experience of losing respect for an idea once they understand better where it comes from.[4]

Another feature of ideas that seems to be significant is that they teach us something. As will be clear from the discussion so far, an important part of what it means to develop as a mathematician is the acquiring of habits of thought that allow certain memories to become generated when previously they might have been directly stored. If, while reading a proof, one comes upon a step that seems arbitrary, and if one then comes to see why it is not arbitrary after all, then one is more likely to be able to think of similar steps for oneself in the future. So perhaps when mathematicians talk of proofs containing ideas [5] what they are referring to is demonstrations of how to generate a step that would otherwise not have sprung to mind.

[4]I cannot resist drawing a parallel with the experience of listening to music. If one is completely unfamiliar with the idiom of a piece, then the piece may well seem too arbitrary to be interesting. At the opposite extreme, if one knows exactly how the composer achieves all his or her effects, then the piece does not deliver enough surprises. The ideal lies somewhere in between, when one understands enough about a piece to feel strongly that there is more to understand.

[5]They really do do this. I once gave a talk after which one of the cleverest mathematicians I know told me solemnly that the talk had contained three ideas.

8 *Mental arithmetic and the concept of width*

Not many mathematicians take mental arithmetic seriously, but, as I hope to show, lessons can be learned from how we go about it. I am not talking about the feats of calculating prodigies but about simpler calculations that are within the grasp of many people. I want to address two questions. The first is whether arithmetical questions have intrinsic features that account for how difficult they are, and the second is why, when we do calculations in our heads, we are often rather unsystematic, finding it easier to think of little ad hoc tricks than to apply a single general procedure. The answers to these questions turn out to be more or less the same.

Consider first the task of adding together two very large numbers, such as 49326394589263498732469689989679 and 239487478799283400698345985538. Obviously, the numbers will be hard to remember, but suppose that they are written down, one above the other with their digits aligned conveniently as follows,

$$\begin{array}{r} 49\,326\,394\,589\,263\,498\,732\,469\,689\,989\,679 \\ +\,239\,487\,478\,799\,283\,400\,698\,345\,985\,538, \end{array}$$

and suppose that the task is to add them up without writing anything down apart from the digits of the sum.

This is not too hard if one does it in the usual way, starting at the end and working backwards, adding the numbers digit by digit and carrying 1 if necessary. The thoughts that would go on in one's mind would be something like these:

> $9+8=17$, *so I'll write down 7 and carry* 1;
> $7+3=10$, $+1$ *makes* 11, *so I'll write down* 1 *and carry* 1;
> $6+1=7$, $+1$ *makes* 8, *so I'll write down* 8 *and not carry* 1;

and so on.

Now consider a different task, that of multiplying 47 by 52 in one's head. Suppose this time that the numbers are not written down. If one does the calculation by long multiplication, then the sequence of thoughts could be like this.

> $4 \times 5 = 20$ *so* $40 \times 50 = 2000$. *I'll keep that in mind for a bit and work out what multiple of 10 to add to it. Now, what were the numbers again? Oh yes,* 47 *and* 52. *So I want to work out* $4\times2+7\times5$, *which is* $8+35$, *which equals* 43. *So I must add* 430 *to what I had before. What was that again? Oh yes,* 2000. *So I've got* 2430. 2430 2430 2430 2430. *Right, what about the final digits? They were, er, 7 and 2, so I need to add* 14 *to that* 2430 *so I get* 2444.

That was obviously much harder than the first calculation, but it can be made easier with appropriate tricks. Here is another sequence of thoughts that one might have.

> *Hmm.* 47×52 *doesn't look particularly nice, but* 47×53 *is OK because it's a difference of two squares that I can calculate easily. So I'll work that out and subtract 47. Here goes:* $47 \times 53 = (50-3) \times (50+3)$, *which equals* $2500-9$,

> *or* 2491. *Subtracting* 47 *from that is a slight bore so I'll subtract* 50 *and add* 3, *getting* 2441 *and then* 2444.

What is it that makes this method easier than the first one but harder, even if one has a good understanding of multiplication, than the addition of two long numbers? The answer is again connected with memory: roughly speaking, the difficulty of a mental calculation is equal to the maximum amount of directly stored information that one is obliged to hold in one's head at any one time. This quantity I shall call the *width* of the calculation.

To show what I mean by this, let me go back to the example of addition. If you are in the middle of adding two large numbers together, and have just finished writing down a digit of the sum, then you can forget everything about what you have done apart from whether you need to carry a 1 to the next digit. Then you move to the next place along, holding this one piece of information in your head while you add the next two digits together. Once you have done that you will need to have in your head two pieces of information: the sum of two single-digit numbers, which has at most two digits, and whether or not you must carry 1. If this is too big a mental burden then you can reduce it by adding the carried 1 to the digit of the first number (if there is a carried 1) and then adding the digit of the second number. If you do it that way then the most you ever have to do is hold one digit in your head and add it to another digit that you can see written down.

Now let us look at the long multiplication. Near the beginning of that calculation one must remember that $50 \times 40 = 2000$ and at the same time remember the numbers 47 and 52. If one is practised at multiplication, then it will be so obvious that the 2000 had to be a multiple of 100 that the main thing to remember will be that this number begins with 2 then 0. So perhaps one should say that the width of this part of the calculation is 6 (measured in digits that have to be directly stored in the mind). Next, one works out that $4 \times 2 = 8$. At this point one can afford to forget about the 4 of 47 and remember merely that it ends with a 7, so one needs to keep in mind that

> *the first number ends in a* 7;
> *the second number is* 52;
> *the 100s part of the product is* 20;
> *part of the tens part is* 8.

There are still 6 digits to keep in mind, though that is not quite all as one must remember their roles too.

The next step is to multiply the 7 by the 5 from 52 and obtain 35, after which one can forget about the 5. So now one must remember that

> *the first number ends in a* 7;
> *The second number ends in a* 2;
> *the hundreds part of the product is* 20;
> *the tens part of the product is* $8 + 35$.

Very briefly the width has gone up to 7. If you want convincing of the relevance of width, you should try the calculation for yourself: this is the place where you may sweat a little. It is a good idea to do the next step quickly, since once one has worked out that $8 + 35 = 43$, one can forget the 8 and the 35 and remember just the 43, which reduces the width back to 6. The next step is to add 2000 to 430 to obtain 2430. Since the 0 has to be there, this really means storing three digits, so the width is now down to 5—those three digits and the 7 and the 2 stored earlier. Multiplying 7 by 2 gives 14, and once one knows this one can forget the 7 and the 2, so the width remains at 5. Adding 2430 to 14 one can do by thinking $2430 + 14 = 2440 + 4 = 2444$, so the width goes down from 5 to 4 and then stays at 4.

Finally, what about the second method of calculating 47×52? Let us write this out more schematically.

$$\begin{aligned} 47 \times 52 &= (50 - 3) \times (50 + 3) - 47 = 2500 - 9 - 47 \\ &= 2491 - 47 = 2491 - 50 + 3 = 2441 + 3 = 2444\,. \end{aligned}$$

If we again ignore the question of how one remembers the roles that the various digits play, then we see that the part of the calculation that requires us to remember most digits is $2491 - 47$, so the width is 6.

Suppose that we had subtracted 47 from 2491 in a more conventional way. That part of the calculation would then have been like this:

$$2491 - 47 = 2480 - 40 + 11 - 7 = 2480 - 40 + 4 = 2440 + 4 = 2444\,.$$

The width at the second stage here is 7, and again this fits one's subjective experience of difficulty very well.

If we make reducing width a priority, then we may be tempted by an alternative calculation:

$$47\times52 = (50-3)\times(50+3)-47 = 2500-9-47 = 2500-(9+47) = 2500-56 = 2444,$$

which has width 5.

A precise discussion of width as a *psychological* concept is quite difficult. Let me list a few of the complications that arise.

(i) If we are finishing with one digit and introducing another at the same time, it is not always clear whether we have to hold them both in mind for a fleeting moment, or whether one replaces the other. For instance, when one adds 9 to 47 and sees instantly that the answer is 56, does what we store in our minds look like

$$[9;4,7] \to [4;1,6] \to [5,6]\ ?$$

Speaking for myself, a picture like

$$[9;4,7] \to [4\uparrow, 7\downarrow] \to [5,6]$$

seems more appropriate. But should I count the instructions to add and subtract 1 as pieces of information I am storing? I have written them as $\uparrow$ and $\downarrow$ to try to convey that what I feel is more like an upwards and downwards 'pressure' that belongs to the hardware of my brain (or perhaps operating system would be a better analogy), which has a little add-9 module in it. Similarly, when I subtract a two-digit number such as 56 from 100 I know that I can work digit by digit: the last digit I subtract from 10 and the rest I subtract from 9. Or I just 'feel' the answer. In this case I see that I'm adding 50+6 to something that's 6 over on the other side of 50, so the numbers 56 and 44 'slot together'. It feels, therefore, as though I can get from 56 to 44 in a width of 2.

(ii) There is a well-known phenomenon, exploited by card memorizers and others, called 'chunking'. If somebody calls out a sequence of digits and asks you to repeat it, then, not surprisingly, the longer the sequence, the harder you will find the task. Experiments have shown that for most adults, the length where sequences become hard to remember is around 7. A few have very unusual memories and can remember much longer sequences. Some, including many dyslexics, find even shorter sequences hard to remember. Nevertheless, there does seem to be something reasonably universal about the number 7.

What happens if instead of digits from 0 to 9 one uses letters of the alphabet? Surprisingly, the length at which it becomes difficult is still 7. The surprise is because the number of sequences of decimal digits is 10^7, whereas the number of sequences of letters of the alphabet is 26^7, which is bigger by a factor of about 800. Therefore, sequences of 7 letters of the alphabet could in principle be used to encode most sequences of up to 10 decimal digits. But the situation is more extreme still. I invite you to see how easily you can memorize—just for a second or two—the following list of six English monosyllables: cow, sky, car, if, pitch, was. You should have had no difficulty whatsoever, especially because I chose to give you six words rather than seven. And yet there are thousands of monosyllabic words in the English language, so one might have thought that an arbitrary sequence of six of them carried the information of well over 20 decimal digits.

Suppose that you wanted to train yourself to remember sequences of up to 14 digits. Here is one way of doing it. You spend a while learning to think of the numbers from 00 to 99 as *individuals* rather than composite objects. With some of them you might be able to think of some strong association—for instance, if your birthday is on 17 April, then you might think of 17 not as 1-7 but as BIRTHDAY! Or rather, you might think of it as the number that has that particularly birthdayish character. If you do that properly, then each of the 100 numbers comes to occupy only one unit of storage space in your brain rather than two, and you can double the number of digits that you can easily remember.

Similarly, if you want to remember sequences of playing cards, one of the first things you need to do is give simple names to each card: you might for instance learn to think 'cat' automatically when you see the queen of clubs—instead of 'queen of clubs'.

Such is the mysterious power of chunking that a sequence of decimal digits can become easier to remember if one reads it out in a different rhythm. For example, if the sequence 2-9-5-2-6-3-7-0-1 is read at a deliberate pace, with precisely the same intonation given to each number, then most people will find that they cannot reproduce all of it. If, however, it is read in chunks of three, with each chunk read quite rapidly—295–263–701—then remembering it suddenly becomes feasible. (I myself find that I start to see the triples as occupying positions on a number line that stretches from 0 to 999, which helps.)

(iii) In the discussion of mental arithmetic above, it was not always easy to decide how many 'units of storage' were taken up by certain numbers. For example, how much brain space do you need to hold the number 10,000 for a few seconds. It has five digits, but one is more likely to think of it as 10^4. Does that mean that it takes up two units? If so, then why is 10^4 easier to remember than 7^4? The problem is that sometimes what one remembers is not so much a *number* as a simple *property*, such as 'is a power of 10'. Since we use a decimal system, that particular property stays in the mind very easily and remembering it does not feel as though it is taking up any width at all.

A more complicated example is the number 1024. If one is familiar with powers of 2, then this number is particularly easy to remember. What is more, there are many different ways of remembering it, of which two are '2^{10}' and 'the power of 2 that is near 1000'.

Just from introspection it is clear that the part of the brain that deals with visual processing also helps with memory, even when the memory itself is not a visual one. It is partly this that makes it possible to remember certain properties of numbers without any apparent effort. For example, when I think of 1024 as the power of 2 near 1000, what I am really doing is looking at my mental picture of the number line (which, rather peculiarly, is not straight, but turns right at certain numbers that have at different times in my past marked the beginning of the unknown) and seeing 1024 just a bit beyond 1000.

Even if, for reasons such as these, it is not easy to assess width in a precise way, one can come up with precise definitions that do it some justice. For instance, as a first attempt one could define the width of an arithmetical problem to be the minimum, over all calculations that solve the problem, of the maximum number of digits that one needs to write in any line of the calculation when it is presented in the form $a_1 = a_2 = \ldots = a_n$. For example, the width of the problem 35×18 is at most 4 (and hence exactly 4) because

$$35 \times 18 = 70 \times 9 = 630.$$

Of course, for a sequence $a_1 = a_2 = \ldots = a_n$ to count as a valid calculation, the individual equalities $a_i = a_{i+1}$ must all be 'single steps': to make the definition of width more precise, one would have to specify what counts as a single step. In the above calculation, I was assuming that it took just one step to get from 7×9 to 63 (because I know my tables), and one step to get from 70×9 to 7×9 with a 0 on the end (because I know the simple rule for multiplying by 10).

A more sophisticated definition of width allows different notation to be used to represent numbers. This, if combined with prior knowledge, can make a difference. If, for example, like many computer scientists you happen to know that

$2^{16} = 65536$ and are asked for the fourth power of 16, then your calculation may well be

$$16^4 = (2^4)^4 = 2^{16} = 65536\,.$$

The final 65536 is hardwired, so does not contribute to the width. Therefore, the width is 3. Similarly, if you are asked what you get when you divide 10! by 8!, then you will see easily that

$$10!/8! = 9 \times 10 = 90$$

and it seems reasonable to say that the width is 2, at least if one takes 10 to be a single thing rather than the composite 1-0.

I borrowed the word 'width' from theoretical computer science, which has a formal notion of the width of an algorithm. Roughly speaking, this means the amount of storage space you need to run the algorithm on a computer, though the precise definition applies to the so-called circuit model of computation, which is mathematically more tractable than a real computer. Because computers have such huge memories these days, much more attention is paid to the length of time an algorithm takes than to its width. For many purposes, this makes good sense, but I believe that artificial intelligence would benefit greatly from a new paradigm of searching for ultra-low-width computations. To do this would be to tie one hand behind one's back deliberately, but one would then be forced to find algorithms that were much more human in the way they worked.

Here I am implicitly distinguishing between permanently stored data—basic facts that one can call up easily—and the temporary storage space needed for a specific instance of a computation. It would be the latter that had to be kept very small. However, reducing the former would also be interesting, as one would need to find ways of generating basic facts rather than storing them. This should lead to a database with a more complicated, human structure.

9 *How does the notion of width apply to proofs?*

Many of the apparently strange features of human memory can be at least partially explained by the hypothesis that we try to reduce width whenever possible. For example, suppose that you wish to make money at blackjack by training yourself to memorize the order of playing cards as they are dealt. Suppose too that you have learned to associate with each card a certain object (possibly animate). This makes it easy to remember sequences of seven cards, but to memorize a whole pack a new technique is needed. A well-known method is to invent *connections* between neighbouring cards, of a kind that will be hard to forget. For example, if one card is 'cat', in your association scheme, and the next one is 'birthday', then you could visualize a cat blowing out the candles on a birthday cake. Or if you needed something a little stronger you could imagine the cat screeching as it accidentally burned itself on the candles (unpleasant, but you wouldn't forget it).

If you need to remember a list and you have not tried this technique, you will be surprised by how well it works. It is not clear why the connections between items can be made easy to remember, but, given that fact, it *is* easy to see why the technique is a good one: it reduces the width of the task dramatically. To remember the sequence of items, all you have to do once the connections are in place is remember the first one. That then generates the second one via the first connection, and then the second generates the third, and so on. So you only ever have to hold one item in your mind at any one time.

As I have already demonstrated, much the same phenomenon occurs when one tries to remember the sequence of steps in certain proofs, except that now the connections between steps are genuine and systematic, rather than spurious and invented. It is tempting to define the width of a proof (relative to a certain mathematical competence and body of knowledge) as the number of steps, or step-generating thoughts, that one has to hold in one's head at any one time. However, as one might expect, this leads to complications, and the complications are greater than those for mental arithmetic.

For a start, when a proof is communicated, it is usually in written form, so one must be clear about exactly what the task of memorization is. If I am reading a proof and want to be sure that it is correct, then I may have to hold things in my head, but much of the storage is done by the paper itself. If in the middle of one part of the proof I read, 'This follows from Lemma 2.1,' and I have completely forgotten what Lemma 2.1 said, it doesn't matter: I can simply turn back and remind myself. (This might be unavoidable if, for instance, the conclusion of Lemma 2.1 was that some rather complicated formula held in a certain context.) Similarly, if I am lecturing the proof, I can establish Lemma 2.1 and then leave a statement of it up on the blackboard rather than holding it in my mind. But suppose that I want to *feel* the validity of the proof rather than checking it. Or, to ask a related but more precise question, suppose that I want to convince somebody else (with a specified level of knowledge and experience) that a proof is valid, without writing anything down. It is sometimes possible to do this, but by no means always. What makes it possible when it is? Width is certainly important here. For example, some mathematics problems have the interesting property of being very hard, until one is given a hint that suddenly makes them very easy. The solution to such a problem, when fully written out, may be quite long, but if all one actually needs to remember, or to communicate to another person, is the hint, then one can have the sensation of grasping it all at once.

Sometimes it is possible to convince somebody very quickly and efficiently that an argument is correct, provided that they are prepared to accept that certain routine calculations produce certain results. In such a case, the calculations resemble empirical evidence: if they have been done by a reputable person, or, better still, several reputable people independently, then one may trust them. Then it seems natural to say that a proof has low width given the outcome of the calculations. Something like this quality is what makes a proof easy to lecture without notes: if the calculations are routine and of low enough width to fit on a blackboard, and if the proof is of low width relative to the calculations, then the proof can be generated by the lecturer.

But suppose one wishes to present a proof in a journal article. Then there is no need to commit anything to memory, and it might seem that width is not particularly important: what is wrong with having to turn back and remind oneself of the statement of Lemma 2.1? And yet there is no doubt that this can be more than a mere inconvenience. If I read a proof that says, 'and therefore $|f|_p \leq c_p \log n$, by (3.2), (3.6), (4.3) and Lemma 2.1' then, even if I manage to check that this line of the proof follows from the three earlier lines combined with Lemma 2.1, I feel that I have lost track of 'what is going on' (another useful phrase when one is evaluating proofs). Why is checking a proof so different from actively following it?

These questions are less puzzling if one thinks of a written proof not as a static object but as a form of *communication*. It is a well-established principle of linguistics that when we have a conversation, or read a book, we do not take in the stream of information purely passively. Rather, we constantly, and largely subconsciously, extrapolate from the information we have so far received and build up expectations about how it will continue. When it does continue, we compare the new information with what we expected, and adjust our further expectations accordingly.[6] (This, too, can be regarded as a width-reducing practice: we don't have to *remember* all that much of what somebody says; rather, it is as though we use the stream of words to make small adjustments to our complicated and highly interconnected web of knowledge about the world.)

Here is a standard piece of advice given to young mathematicians: when you are reading a paper, do so as actively as possible; don't just accept what the author presents to you; try to guess as much as you can about how the argument will go; try to prove the theorem for yourself; turn to the paper only for the occasional hint when you get stuck. In brief, read the paper as though you were having a conversation with the author. One advantage of a low-width proof is that it makes this advice much easier to follow. A badly written paper will present the complicated statement of Lemma 2.1 with no hint about how it is going to be used later. What expectation does this set up in the mind of the reader? Merely that the author is unlikely to have gone to the trouble of proving the lemma if it was not at some point going to be used. But this will not be much of an expectation if the lemma is difficult to remember, and still less of one if there are other lemmas to hold in mind as well. Unless one has managed to work out the structure of the proof for oneself, which is not possible unless one has considerable expertise in the area, one is faced with an unpalatable choice: either directly memorize a number of arbitrary-seeming statements, or wait for the author to use them at unexpected times later in the proof and refer back to them. Neither of these feels much like conversation.

[6]This is particularly clear when we listen to music. To elaborate on an earlier footnote, a truly satisfying piece is one that is predictable enough to set up expectations, and unpredictable enough not to confirm them too reliably.

A second complication is that the same proof can be presented in many different ways. Is width a property of proofs or of presentations of proofs? The answer is undoubtedly the latter: one of the features of a well-presented proof is that it becomes easier to remember because 'the main ideas are clearly explained'. If my basic thesis is correct, then a major reason both for the clarity and for our liking of the notion of 'main ideas' is that they enable us to memorize an argument more easily because they provide us with a low-width presentation of it. Of course, this raises the difficult question that I have already mentioned: when do two arguments represent different proofs, and when are they merely different presentations of the same proof? (If one had a good answer to this question, then one could perhaps define the width of a proof to be the smallest width of any presentation of it.)

In this chapter, I have not attempted to provide a fully worked-out theory of the width of proofs; rather, I have tried to show that, despite the complications discussed above, it is likely that many of our evaluative words about proofs will be illuminated if we pay proper attention to how we memorize those proofs, and in particular to how many distinct 'arbitrary items' we need to hold in our head when we do so. For many purposes, even a rather vague concept will do. For instance, there are often circumstances where it is obvious that, whatever width might mean, presentation A of some proof has lower width than presentation B. However, I want to stress that I believe that with more work it would be possible to make the notion of width much more precise, even if the precision fell short of a formal definition. In particular, I envisage a concept that is more precise than subjective-sounding concepts such as 'transparent', or even 'easily memorable'. Low width is supposed to be a lower-level feature of presentations that *makes* them transparent or easily memorable.

One of my reasons for paying attention to presentations is a probably forlorn hope that it could encourage mathematicians to write better papers. Anybody with any experience of mathematics, particularly at university level, knows that non-conversational presentations of the kind I have just discussed are everywhere—in lecture courses, textbooks, and research papers. How can one combat this? It is obviously not enough to complain about the situation in general, but complaining about specific instances of bad mathematical writing is rude and will simply antagonize people. In any case, some mathematicians freely admit to writing badly, but they don't have a clear idea how they could improve—they declare that they are not 'good expositors' and leave it at that.

A possible solution to the problem is to identify objective features of proofs that are clearly desirable, and use emotionally neutral words to describe them: it will not do to describe a proof as 'badly written' or 'unclear,' however objective an assessment that may be on occasion. But neutral features, such as length, for example, do not seem to cause problems: all other things being equal, a short proof is preferable to a long one, and if you show somebody how to shorten one of their proofs, they will usually be grateful to you. I would like width to be given as much, if not more, attention than length. It may not be the answer to everything, but there are many papers that give high-width presentations of proofs, when low-width presentations would be possible and far easier to read.

10 Concluding remarks

When I was invited to submit a written version of my talk to these conference proceedings, I readily agreed, but by the time I got round to doing any writing I found, to my initial dismay, that I had lost my transparencies. However, it turned out that I had given a low-width talk (relative to my particular expertise), and I was therefore able to generate it all over again. I was able to check this because when I had almost finished I found the transparencies again.

I hope that this chapter has demonstrated just how much we do not fully understand about the nature of proofs, and how many interesting philosophical questions arise when we try to. I hope too that it gives an idea of the form that such understanding might take, even if it only scratches the surface of a huge subject.

Let me finish with the answer that I promised to give to the question about repetitive numbers. If you are a mathematician then it is a great pity if you read on: this is your last chance to avert your eyes and solve the problem for yourself. I will not in fact give a full answer, but just the following hint: can you find a number that is not repetitive?

Is there a problem of induction for mathematics?

ALAN BAKER

1 Introduction

'Induction' is a term which means one thing in the context of mathematics and quite another in the context of philosophy. In mathematics, induction is a familiar (and highly useful) method of proof. To show that a conjecture, $C(n)$, holds for all natural numbers, it suffices to show that it holds for $C(1)$—the so-called base step—and that if it holds for $C(m)$ then it holds for $C(m+1)$—the induction step. Mathematical induction of this sort is straightforwardly deductive. In philosophy a distinction is standardly made between deductive and non-deductive methods of rational support, and these latter methods (which may include inference to the best explanation, abduction, analogical reasoning, etc.) are often referred to collectively as 'inductive reasoning', and studied using 'inductive logic'. Clearly mathematical induction is not 'inductive' in this broader philosophical sense. Induction in the narrow mathematical sense is an important—indeed indispensable—mathematical tool, and its use is almost entirely uncontroversial. Induction in the broad philosophical sense is a large and amorphous topic, and its application in the domain of mathematical reasoning has been addressed in some detail by Polya (1954).

My title question, however, uses the term 'induction' in a third sense. I am interested specifically in the non-deductive form of reasoning known as *enumerative induction*. A (hackneyed) example of this sort of reasoning is the following:

Emerald E_1 is observed to be green.
Emerald E_2 is observed to be green.
- - - - - - -
Emerald E_n is observed to be green.
___
Hence, all emeralds are green.

My primary aim in this paper is to sketch preliminary answers to the following two questions:

(A) Does the mathematical community ever rely on enumerative induction to underpin its belief in a mathematical claim?
(B) Ought enumerative inductive reasoning to ever justify our belief in a mathematical claim?

For the purposes of the project of this chapter I shall be making a couple of background assumptions. First, I am assuming that we do know plenty of mathematics, and that deduction from axioms is a primary mechanism for the acquisition of such knowledge. Thus in addressing the factual question, (A), I take the issue to be not whether (à la Mill) *all* mathematical knowledge is inductive, but whether *any* mathematical knowledge is inductive. Second, I am adopting a broadly naturalistic stance which takes seriously the actual patterns of reasoning and epistemic attitudes of mathematicians. I take the project of answering the normative question, (B), to involve steering a middle course between radical scepticism on the one hand and uncritical acceptance of actual mathematical practice on the other.

I shall also be focusing attention specifically on number theory. Since enumerative induction paradigmatically ranges over denumerable domains, it is not surprising that it is claims restricted to the natural numbers which tend to feature in inductive arguments of this form. Nonetheless it is worth bearing in mind that the conclusions I defend do not necessarily apply to other areas of mathematics, should it turn out that enumerative inductive arguments feature in such areas.

The structure of the chapter is as follows. In sections 1 and 2, I investigate the empirical question, (A), and conclude that there is prima facie evidence for enumerative induction playing a justificatory role in number theory. In particular there are cases in which more positive instances lead to more confidence in the truth of a conjecture. In sections 3 and 4, I turn to the normative question, (B), and argue that there is a tension here with the answer given to (A) because there are strong philosophical grounds for doubting the legitimacy of enumerative induction over the domain of the natural numbers. In section 5, I propose a solution which reconciles the two answers by showing that the evidence for enumerative induction playing a justificatory role is flawed, and that other explanations can be given for mathematicians' confidence in the truth of the relevant conjectures. I conclude that ultimately there is no problem of induction for mathematics.

2 *Enumerative induction and discovery*

We shall begin with two historical examples in which enumerative induction appears to have played a pivotal role. In 1650, Pierre de Fermat conjectured that every number of the form $F_n = 2^{(2*n)} + 1$ is prime. This conjecture seems to have been based purely on enumerative induction from the first five such numbers (now known as Fermat numbers)

$$F_0 = 3$$
$$F_1 = 5$$
$$F_2 = 17$$
$$F_3 = 257$$
$$F_4 = 65\,537$$

all of which Fermat had shown to be prime. It was not until more than 100 years later that Euler showed that the next case, $F_5 = 2^{32} + 1 = 4\,294\,967\,297$, is divisible by 641 and so is not prime. At present, only composite Fermat numbers are known for $n > 4$ and it seems unlikely that any more prime Fermat numbers will be found using current computational methods and hardware. Thus Fermat's inductive conjecture could scarcely have been more wrong: what he conjectured to be a universal property of numbers of this form turns out to perhaps be unique to the first four cases!

The second historical example concerns perfect numbers, in other words numbers which are equal to the sum of their divisors (including 1, but excluding themselves). On the basis of knowing only the first four perfect numbers 6, 28, 496, 8128 the ancient Greeks conjectured, inductively but incorrectly, that

(i) the nth perfect number contains exactly n digits
(ii) the even perfect numbers end, alternately, in 6 and 8.

On the other hand, the Greeks also conjectured the following, both of which are currently open.[1]

(iii) All perfect numbers are even
(iv) All perfect numbers are of the form $2^{k-1}(2^k - 1)$, where $2^k - 1$ is prime.

If we tally up the score card for our five examples of enumerative induction from the history of mathematics, we are left with three definite failures and two undecided—hardly a confidence-inspiring track record! However it is important, I think, to distinguish here between context of discovery and context of justification. The primary role of enumerative induction in the above cases was to suggest conjectures that are worth considering and worth trying to prove or refute by other means. The (meagre) inductive evidence was not considered as justifying belief in the truth of these conjectures.

3 *The descriptive question: Two case studies*

Do things get any better when we shift our attention to more recent mathematics? In order to help answer this question, I want to look in some detail at two different conjectures—Goldbach's Conjecture (GC), and the Even Perfect Number Conjecture (EP)—for which the current evidence seems to be based primarily on enumerative induction.

[1]These, together with (i) and (ii), are listed as conjectures by Nicomachus of Gerasa in his *Introductio Arithmetica* of c. AD100. A fifth conjecture, that there are infinitely many perfect numbers, is also still an open question in number theory. However it is less clear whether this conjecture is inductive in nature.

Goldbach's Conjecture

In a letter to Euler written in 1742, Christian Goldbach conjectured that all even numbers greater than 2 are expressible as the sum of two primes.[2] Over the following two and a half centuries, mathematicians have been unable to prove GC. However it has been verified for many billions of examples, and there appears to be a consensus among mathematicians that the conjecture is most likely true. Let us begin by examining these three claims in a little more detail.

Below is a partial list (as of October 2003) showing the order of magnitude up to which all even numbers have been checked and shown to conform to GC.

Bound	Date	Author
1×10^3	1742	Euler
1×10^4	1885	Desboves
1×10^5	1938	Pipping
1×10^8	1965	Stein & Stein
2×10^{10}	1989	Granville
1×10^{14}	1998	Deshouillers
6×10^{16}	2003	Oliveira & Silva

Despite this vast accumulation of individual positive instances of GC, aided since the early 1960s by the introduction—and subsequent rapid increases in speed—of the digital computer, no proof of GC has yet been found. Not only this, but few number theorists are optimistic that there is any proof in the offing. Fields medallist Alan Baker (no relation) stated in a 2000 interview, 'It is unlikely that we will get any further [in proving GC] without a big breakthrough. Unfortunately there is no such big idea on the horizon.'[3] Also in 2000, publishers Faber & Faber offered a $1 000 000 prize to anyone who proved GC between 20 March 2000 and 20 March 2002, confident that their money was relatively safe.

What makes this situation especially interesting is that mathematicians have long been confident in the truth of GC. Hardy and Littlewood asserted, back in 1922, that 'there is no reasonable doubt that the theorem is correct', and Echeverria, in a recent survey article, writes that 'the certainty of mathematicians about the truth of GC is complete' (Echeverria 1996: 42). Moreover this confidence in the truth of GC is typically linked explicitly to the inductive evidence: for instance, G. H. Hardy described the numerical evidence supporting the truth of GC as 'overwhelming'. In the light of such confidence, it seems prima facie reasonable to conclude that the grounds for mathematicians' belief in GC is the enumerative inductive evidence.

[2]In fact, Goldbach made a slightly more complicated conjecture which has this as one of its consequences.

[3]*The Times*, 16 Mar. 2000.

The Even Perfect Number Conjecture

The second case study I shall consider is the conjecture, which as previously mentioned dates back at least to Nicomachus in AD100, that all perfect numbers are even. The Greeks knew of four perfect numbers. Today we have discovered around forty perfect numbers, and all of these are even. Below is a partial list, showing the order of magnitude of the largest known perfect number.

Date	Number of perfect numbers	Size of largest perfect number	Author
300BC	4	8×10^3	Euclid
1536	5	3×10^7	Regius
1603	7	1×10^{11}	Cataldi
1738	8	2×10^{18}	Euler
1911		$2^{88}(2^{99} - 1)$	
2001	39	$1 \times 10^{4\,000\,000}$	

As with Goldbach's Conjecture, this inductive evidence is important because no deductive proof of EP has yet been found. However, mathematicians' opinions about the truth or falsity of EP are considerably less settled than for GC. Here are two sample quotations from recent sources:

The existence of odd perfect numbers appear[s] unlikely. (Guy 1994: 44–5)

The issue of odd perfect numbers remains unsettled, however. No one knows whether there are any. (*Science News*, 25 Jan. 1997)

The key information from the above two case studies is summarized in the following table.

Conjecture	Date conjectured	Number of verified cases	Mathematicians' degree of belief
GC	1742	6×10^{16}	definitely true
EP	c. AD100	39	no consensus

Mathematicians' degree of belief in each conjecture is clearly correlated with the strength of its enumerative inductive support. Of course reference to a mere two examples is itself a shaky inductive basis from which to make any general claims about the justificatory role of enumerative induction in number theory, let alone in mathematics more generally. Nonetheless it seems as if our answer—at least tentatively—to question (A) should be yes. In at least some cases, mathematicians do make use of enumerative induction to underpin their belief in the truth of certain mathematical claims.

4 *Hume's problem of induction*

I want to approach the normative question, (B), concerning the rational justification of induction in mathematics via the broader question of what features make the mathematical case of enumerative induction distinctive from the empirical case. In a sense, therefore, I am adding a third question to the mix.

(C) Is the use of induction in mathematics more or less rationally justified than its use in the empirical case?

Looming large in any discussion of induction in the empirical case is the so-called problem of induction which finds its classic expression in the writings of David Hume. Hume's original question concerned how we come to know about unobserved matters of fact. His notorious conclusion was that, although we cannot help but reason inductively, such reasoning is in an important sense not rationally justifiable. In particular, the only way to make inductive reasoning secure is by appeal to some sort of 'Principle of Uniformity'. But such a Principle is itself only justifiable inductively, and there is no escape from the inductive circle.

The precise 'solution' to Hume's problem of induction is not agreed upon, but that there is a solution is generally conceded. For present purposes, therefore, I shall simply assume that we do have good rational grounds for trusting inductive inference in the empirical case. Thus we shall be content if it turns out, in response to the normative question (B), that mathematics is at least as comfortably off as empirical science with regard to the security of its inductive methods.

A couple of reasons for thinking that the problem of induction ought to be less pressing for mathematics is that enumerative induction is not widely used in mathematics (*pace* Mill) and that when it is used it is more often as a method of discovery rather than of justification. Both these claims are plausible, but even if they are true this only affects the *scope* of the problem, not its severity. In Goldbach's Conjecture and the Even Perfect Number Conjecture we have two prima facie cases of enumerative induction used in the context of justifying a mathematical claim. If these cases are genuine, then the problem of induction in mathematics must be faced. It is also worth mentioning that even in cases where the use of enumerative induction is restricted to the context of discovery, there is still a potential problem of induction that remains (although not Hume's problem). For even in using enumerative induction to discover conjectures worth attempting to prove, we need to have an idea of which mathematical properties are projectable, and so Goodman's new riddle looms large here.

A third reason for dismissing the problem is that in mathematics, unlike in empirical science, deductive methods (especially deductive proof) are always in principle available. In other words, enumerative induction is dispensable as a justificatory mathematical tool. However, it is not clear that the premise of this argument is true, for we know from Gödel's results that, given a fixed set of axioms, there are true number-theoretic claims which are not deductively provable from these axioms. It is conceivable, therefore, that Goldbach's Conjecture is undecidable relative to the standard axioms of Peano Arithmetic. Note that if this is the case then the Conjecture is true, since if it is false then there is some even number for which it fails, and this failure can be derived in a finite number of steps. If this were the situation—and we have no definitive reason to think that it is not—then our only way of justifying the truth of GC might be via inductive rather than deductive means.

I conclude that none of these reasons provides strong grounds for dismissing the problem of induction for mathematics out of hand. Moreover, there are at least a couple of countervailing reasons for thinking that an investigation of enumerative induction in mathematics might be more rewarding than further investigation of the empirical case. First, the mathematical version of induction has been much less investigated and discussed by philosophers than its empirical counterpart. Second, there does seem to be genuine disagreement—both among mathematicians and among philosophers—concerning whether such inductive evidence does provide good grounds for belief in the truth of a mathematical conjecture. Thus there is room to raise epistemological questions in this context without necessarily flying in the face of common sense.

5 *The normative question: Is enumerative induction in mathematics rationally justified?*

That there are significant differences between cases of enumerative induction in science and in mathematics is undeniable. The issue is whether any of these differences make a difference to whether—and if so, how much—any conclusion extracted from such inductions might be rationally justified.

Let me begin by surveying three distinctive features of the mathematical case which I claim do not (or ought not to) make a difference. First, individual instances of a universal mathematical hypothesis are typically provably true (or false). Thus in the case of GC, any given even number can be checked in a finite number of steps to verify whether it can be expressed as the sum of two primes. This feature is often stressed in philosophical analyses of induction in mathematics (see, e.g., van Bendegem 1998), but it is not clear how—if at all—it is relevant to the logical relation between individual instances and the universal hypothesis under which they fall.[4] Second, the mathematical hypothesis under which the various instances fall might itself be provable. If so then there is a sense in which justification by means of enumerative induction would be redundant, at least in principle. A couple of considerations are relevant here. For one thing, redundancy in principle is very different from redundancy in practice. Even if a proof exists, there is no guarantee that a proof of a size or complexity accessible to the human mind is out there to be found. Moreover, Gödel's Incompleteness Theorem shows that there is no guarantee that a given hypothesis is provable even if

[4]The provability of individual instance does, however, cause problems for probabilistic analyses of the use of enumerative induction in mathematics.

it is true.[5] And even if a feasible proof is possible, it is far from obvious why this should undermine alternative, inductive modes of justification of the same result. A third distinctive feature is that the domain across which a given mathematical hypothesis ranges may be known to be infinite, as is the case for GC. Empirical hypotheses do not typically have this feature. Thus the hypothesis that all emeralds are green, while certainly open-ended, does not necessarily have an infinite number of instances.

The one distinctive feature of the mathematical case which ought to make a difference to the justification of enumerative induction (or so I shall argue) is the importance of order. By this I mean that the instances falling under a given mathematical hypothesis (at least in number theory) are intrinsically ordered, and furthermore that position in this order can make a crucial difference to the mathematical properties involved. My approach here is inspired in part by remarks made by Frege in Section 10 of his *Grundlagen*. Frege was interested in the stronger (and less plausible) claim of Mill that *all* knowledge of mathematical truths is essentially empirical and inductive in nature, whereas our current concern is more narrowly focused on whether induction is *ever* justified within the context of mathematics. Frege writes, with regard to mathematics, that

> the ground [is] unfavourable for induction; for here there is none of that uniformity which in other fields can give the method a high degree of reliability.

He then goes on to quote Leibniz, who argues that difference in magnitude leads to all sorts of other relevant differences between the numbers.

> An even number can be divided into two equal parts, an odd number cannot; three and six are triangular numbers, four and nine are squares, eight is a cube, and so on.

Frege also explicitly compares the mathematical and non-mathematical contexts for induction.

> In ordinary inductions we often make good use of the proposition that every position in space and every moment in time is as good in itself as every other. Position in the number series is not a matter of indifference like position in space.

Taking our cue from Frege's remarks, one way to underpin an argument against the use of enumerative induction in mathematics is via some sort of *non-uniformity principle*: in the absence of proof, we should not expect numbers (in general) to share any interesting properties. (It would be analogous, perhaps, to changing the atomic arrangement of an element and expecting the new element to have significant chemical properties in common with the original.) Hence establishing that a property holds for some particular number gives no reason to think that a second, arbitrarily chosen number will also have that property.[6] Rather

[5]In cases of enumerative induction the logical form of the hypothesis is always universal. If, for example, GC is false, then there is some finite even number which is a counterexample. Hence it is provably false. So since falsity entails decidability, the only way GC could be undecidable is if it were true.

[6]Frege: it is difficult to find even a single common property which has not actually to be first proved common.

than the Uniformity Principle, which Hume suggests is the only way to ground induction, we have almost precisely the opposite principle! It would seem to follow from this principle that enumerative induction is unjustified, since we should not expect (finite) samples from the totality of natural numbers to be indicative of universal properties. But perhaps the apparent weakness here—of general non-uniformity—can be turned around to provide an argument *in favour* of enumerative mathematical induction, as follows. When some property does turn out to be shared by a large array of numbers, with no known exceptions, the best explanation of this pattern is that the property holds universally. In other words, since interesting common properties of a succession of different numbers are rare, when such a property is found in some long sequence, then the best explanation is that there is some theorem (or at least universal truth) which entails it. Consider the following example. I take some number, say 641, and discover that the sum of the squares of its numerals is divisible by 53. This should give me no particular reason to expect another randomly chosen number to have this property. But if I start picking other numbers, and they all have this property, this seems to provide an inference to the best explanation-style argument for this being a property of *all* numbers. (All numbers have property P is the simplest/best explanation of the series of otherwise surprising results.)

I think that this response fails, and that the reason for its failure actually reveals the underlying problem with enumerative induction in mathematics. The problem, in the case of GC and in all other cases of induction in mathematics, is that the sample we are looking at is *biased.*

Note first that *all* known instances of GC (and indeed all instances it is possible to know) are—in an important sense—small. In a very real sense, there are no large numbers: Any explicit integer can be said to be small. Indeed, no matter how many digits or towers of exponents you write down, there are only finitely many natural numbers smaller than your candidate, and infinitely many that are larger.

Of course, it would be wrong to simply complain that all instances of GC are *finite*. After all, every number is finite, so if GC holds for all finite numbers then GC holds *simpliciter*.[7] But we can isolate a more extreme sense of smallness, which I shall call *minuteness.*

Definition: a positive integer, n, is *minute* just in case n is within the range of numbers we can (given our actual physical and mental capabilities) write down using ordinary decimal notation, including (non-iterated) exponentiation.

Verified instances of GC to date are not just small, they are minute. And minuteness, though admittedly rather vaguely defined, is known to make a difference. Consider, for example, the logarithmic estimate of prime density which is exceeded only at a huge bound. If the Riemann Hypothesis is true, then an upper bound on its size (the first Skewes number) is 8×10^{370}. Though an impressively large number, it is nonetheless minute according to the above definition.

[7] Cf. Wang's Paradox (discussed, for example, by Dummett (1978)).

However if the Riemann Hypothesis is false then an upper bound at which the logarithmic estimate is first exceeded (the second Skewes number) is $10 \uparrow 10 \uparrow 10 \uparrow 10 \uparrow 3$.[8] The necessity of inventing an arrow notation here to represent this number tells us that it is not minute. The second part of this result, therefore, although admittedly conditional on a result that is considered unlikely (viz. the falsity of RH), implies that there is a property which holds of all minute numbers but does not hold for all numbers. Minuteness can make a difference.

Hence the sample of positive instances of GC is biased, and unavoidably so.[9] Imagine, for example, that mathematicians had only looked at even numbers divisible by 4 when checking GC, or only (even) square numbers. Presumably such evidence would carry less weight since the range of instances is comparatively unvaried.

A defender of induction in mathematics might respond that matters are no worse than in the empirical case. There are many distinctive features which are common to all observed emeralds, ravens, electrons, and so on; for example, they have all been observed before the present, and they are all within the past light cone of the Earth. So why not argue, on analogous grounds, that empirical induction is biased? The disanalogy, as already mentioned, is that the position of a number in the ordering of integers often does make a difference to its mathematical properties. There are no corresponding systematic differences between past and future or between inside and outside the Earth's light cone. Indeed, insofar as there are any general theoretical principles they tend to concern the spatial and temporal invariance—other things being equal—of fundamental physical properties. Of course there is still room for a purely sceptical worry concerning induction in the empirical case, but it seems to lack the specific motivation for worry which afflicts induction in mathematics.

6 *Re-examining the descriptive question*

Now if enumerative induction in the mathematical context is really as problematic as I have suggested, then this puts pressure on our original answer to the descriptive question. For our survey of GC and EP in section 2 seemed to indicate that mathematicians *do* use enumerative induction. Thus we are faced with the issue of explaining the difference between the almost universally believed GC, which has extensive enumerative inductive support, and the shaky EP, which has sparse enumerative inductive support.

One response would be to claim that mathematicians are simply being *irrational*. This is compatible with the broadly naturalistic attitude I espouse (philosophy could be right where all the mathematicians are wrong), but it is not a

[8] Here $\uparrow$ denotes the exponentiation function, and is evaluated from right to left. For more on this example, see te Riele (1987).

[9] Unavoidable, unless some way of checking non-minute numbers for conformity to GC can be found. This seems unlikely given that it would require testing numbers of the same (non-minute) order of magnitude for primality, which is a comparatively complex computational task.

comfortable position. My response, by contrast, will be to argue that the connection between enumerative inductive evidence and mathematicians' beliefs is only apparent. In other words, I want to reverse our preliminary judgment about the answer to question (A): enumerative induction does not play the role it seems to in these cases. For ease of exposition, I shall treat the two cases of GC and EP separately since I think that the considerations involved are quite distinct.

Goldbach's Conjecture

Echeverria discusses the important role played by Cantor's publication, in 1894, of a table of values of the Goldbach partition function, $G(n)$, for $n = 2$ to 1,000. (Echeverria 1996: 29-30) The partition function measures the number of distinct ways in which a given (even) number can be expressed as the sum of two primes. Thus $G(4) = 1$, $G(6) = 1$, $G(8) = 1$, $G(10) = 2$, etc. This shift of focus onto the partition function coincided with a dramatic increase in mathematicians' confidence in GC. However, Cantor did not simply provide more of the same sort of inductive evidence, since Desboves had already published, in 1855, tables verifying GC up to 10000. To understand why Cantor's work had such an effect it is helpful to look at the following graph which plots values of the partition function, $G(n)$, from 4 to 100000.[10] This graph makes manifest the close link between $G(n)$ and increasing size of n. Note that what GC amounts to in this context is that $G(n)$ never takes the value 0 (for any even n greater than 2). The overwhelming *impression* made by the above graph is that it is highly unlikely for GC to fail for some large n. At the upper end of this graph, for numbers on the order of 100,000, there are always at least 500 distinct ways to express each even number as the sum of two primes!

However, as it stands this graph is purely heuristic. The roughly thirty years following Cantor's publication of his table of values (described by Echeverria as the second period of research into GC) saw numerous attempts to find an analytic expression for $G(n)$. If this could be done then it would presumably be comparatively straightforward to prove that this analytic function never takes the value 0 (Echeverria 1996: 31). By around 1921, pessimism about the chances of finding such an expression led to a change of emphasis, and mathematicians started directing their attention to trying to find lower bounds for $G(n)$. This too has proved unsuccessful, at least to date. Thus consideration of the partition function has not brought a proof of GC any closer. However it does allow us to give an interesting twist to the argument of the previous section. The graph suggests that the hardest test cases for GC are likely to occur among the smallest numbers; hence the inductive sample for GC *is* biased, but it is biased *against* the chances of GC. This insight allows the tension between our answers to the normative question and to the descriptive question to be defused, at least in the particular case

[10]Of course the number of results displayed here is orders of magnitude beyond Cantor's own efforts, but the qualitative impression is analogous. This graph is taken from Mark Herkommer's Goldbach Conjecture Research website at http://www.petrospec-technologies.com/Herkommer/goldbach.htm (last visited 3 June 2007).

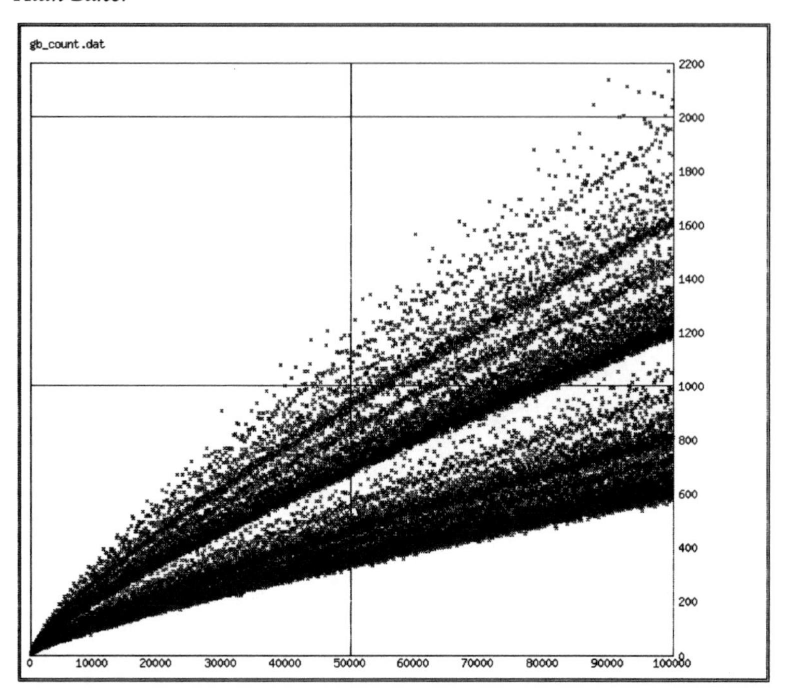

of GC. For mathematicians' confidence in the truth of GC is not based purely on enumerative induction. The values taken by the partition function indicate that the sample of positive instances of GC is indeed biased, and biased samples do not—as a general rule—lend much support to an hypothesis. But in this particular case the nature of the bias makes the evidence stronger, not weaker. So it is possible to argue that enumerative induction is unjustified while simultaneously agreeing that mathematicians are rational to believe GC on the basis of the available evidence.

Note that gathering data simply about whether each successive even number conforms or fails to conform to GC does not in itself yield any information about small number bias. Such data would be closer to pure enumerative induction for GC. The historical situation before Cantor's work at the end of the nineteenth century indicates that more of this same sort of basic evidence would not have yielded such a strong consensus about the truth of GC.

A prima facie strong objection against this approach to reconciling the normative and descriptive aspects of enumerative induction is that the evidence provided by the graph of the partition function falls short of *proving* any positive bias. In the absence of a known function corresponding to $G(n)$, or a lower bound for it, the evidence in favour of $G(n)$ never being 0 is essentially just more inductive evidence. (Moreover, $G(n)$ is not monotonically increasing even within the range of small numbers studied.)

One response to this objection is to complain that it misleadingly conflates two importantly distinct senses of induction. The evidence encapsulated in the graph is inductive in the sense of not deductively proving anything about GC, but it is not inductive in the narrow sense of enumerative induction with which we are primarily concerned. Rather it is a more sophisticated version of what might be termed functional induction. Instead of there being just some basic property (expressibility as the sum of two primes, colour, etc.) which is either present or absent, there is a numerical (functional) relation which can take infinitely many different values for different n.

Another response is to try to bolster the evidence for GC via other auxiliary arguments. One candidate is a heuristic probabilistic argument based on the distribution of primes (O'Bryant n.d.: 3). The number of primes tends to $n/\log n$ (as n increases), from the Prime Number Theorem (which asserts that the density of primes around $n \sim 1/\log n$). Hence there are approximately $\sim n^2/log^2 n$ sums of primes, each of which is less than $2n$. Hence a typical integer less than $2n$ can be written as a sum of 2 primes in $n/\log^2 n$ ways. GC asserts that the partition function is never 0. Meanwhile this probabilistic approximation to the function tends rapidly to infinity! Of course, there is no conclusive reason to think that the primes *are* randomly distributed. But while this may be a heuristic argument, it does not seem to depend on enumerative induction in any substantive way.[11]

The Even Perfect Number Conjecture

The descriptive data pertaining to the Even Perfect Number conjecture (EP) poses less of a problem for our normative conclusion than does the corresponding data for GC. For since there seems to be little or no consensus concerning the truth of EP, this is consistent with enumerative induction playing no justificatory role. Nonetheless, I think that it is worth returning to consider EP because it provides a useful context for analysing further the notion of positive instance in the mathematical context.

In the figure on page 63 it was noted that EP has to date around 39 positive instances, in other words 39 perfect numbers (all even) have so far been discovered. This comparative paucity of instances seemed to fit well with the uncertainty over the truth of the general EP conjecture under which they fall. But notice that—from another perspective—EP has huge numbers of positive instances. For EP is logically equivalent to the following Odd Imperfect Number Conjecture (OIN).

(OIN) All odd numbers are imperfect.

Now OIN almost immediately yields more positive instances than does EP; after all there are more than 39 odd numbers less than 100 which have all been checked to be imperfect. Not only this, but there are some strong lower-bound

[11] Another line of auxiliary argument might be based on the various partial results relating to GC. In 1931, Schnirelmann proved that every even number can be written as the sum of not more than 300,000 primes(!). This upper bound on the number of primes required has since been lowered to 6 (Ramaré 1995). In addition, Chen (1978) proved that all sufficiently large even numbers are the sum of a prime and the product of two primes. Such results do not seem to make the truth of GC any more likely. But perhaps they provide evidence that GC is provable.

results on the size and composition of odd perfect numbers.[12] For example, Brent et al. (1991) proved that any odd perfect number must have at least 8 distinct factors and at least 300 digits. Also, Sayers (1986) proved that any odd perfect number must have at least 29 prime factors (not necessarily distinct), and Brandstein (1982) proved that it must have some prime factor greater than 5×10^5.

Since it has been proved that any odd perfect number is greater than 10^{300}, we have ruled out almost 10^{300} potential falsifiers of OIN, and hence of EP. So in one sense the enumerative inductive evidence for EP is much *more* extensive (by a factor of about 10^{285}) than that for GC! My argument is obviously inspired by Hempel's well-known objection to Nicod's criterion of confirmation, that any generalization of the form All *A*s are *B*s is confirmed by the observation of an object which is both *A* and *B*. Hempel points out that—assuming confirmation is preserved by logical equivalence—this entails that 'All ravens are black' is confirmed by the observation of a white shoe. One objection to Hempel's manoeuvre is that the logically equivalent hypothesis, all non-black things are non-ravens, makes use of predicates which do not correspond to natural kinds. Hence it is not a legitimate scientific hypothesis. Whether some analogous notion of natural kind can be carried over to the mathematical case is an interesting question (and one which has been little explored by philosophers). Even if it can, it does not seem as if the predicates appearing in OIN are any less natural than those in EP. Even and odd are precisely symmetrical, and while imperfect might seem to be purely negatively defined (and thus to share the problematic features of non-raven and non-black), it is far from clear that perfect is itself a particularly natural predicate. Aside from its historical links to ancient Greek numerology, it is hard to see why the property of being equal to the sum of all proper divisors including 1 should be deemed of any special mathematical significance.

To the extent that the line of argument I sketched above has force, it lends support to the normative conclusion reached at the end of Part 4. If it is correct to think of EP has having huge numbers of positive instances, then the fact that there is nonetheless no clear agreement that it is true indicates that mathematicians put little stock in enumerative induction, and that they are right not to do so.

7 *Conclusions*

I conclude that mathematicians ought not to—and in general do not—give weight to enumerative induction *per se* in the justification of mathematical claims. (To what extent enumerative induction plays a role in the discovery of new hypotheses, or in the choice of what open problems mathematicians decide to work on, is a separate issue which I have not attempted to address here.) More precisely, my thesis is in two parts:

[12]There is a sense in which writing a mathematics paper about the properties of odd perfect numbers is akin to writing a biology paper about unicorns.

(i) Enumerative induction **ought not** to increase confidence in universal mathematical generalizations (over an infinite domain);
(ii) Enumerative induction **does not** (in general) lead mathematicians to be more confident in the truth of the conclusion of such generalizations.

If this is right, then verifying ever larger even numbers should not (and does not) increase mathematicians' confidence in GC. Why then do they keep on with such verifications? There are several potential reasons. First, it is a clearly defined and concrete task, and one which generates a certain amount of publicity for whoever holds the current record (compare, for example, the quest for ever more gargantuan primes). Second, the task provides a convenient test bed (and showcase) for advances in computer processing speed and algorithmic programming efficiency. Third, in the process of checking cases, data may be gathered to provide a basis for more sophisticated forms of evidential reasoning, such as values for the partition function. Induction, at least in the narrow, enumerative sense, *is* more problematic in mathematics. However, although I am admittedly myself inducing from just two case studies, my claim is that mathematicians do not base knowledge claims on enumerative inductive evidence alone. Hence the answer to my title question, asking whether there is a problem of induction for mathematics is yes in principle, but no in practice.

The cognitive basis of mathematical knowledge

MARINELLA CAPPELLETTI AND VALERIA GIARDINO

In 1988 Nicolas Slonimsky, aged 94, the world-renowned lexicographer of music and a conductor, created a new composition by identifying each note in Beethoven's 5th Symphony with a number and then playing the square root of each note. (Slonimsky and Bonotto 2002)

What is involved in our understanding of numbers such as those contained in the above sentences? Are we born with this ability or do we learn it in life? Can we lose our capacity to understand and use numbers following a brain injury?

The present chapter addresses the above questions from a cognitive perspective. We discuss how cognitive science can contribute to the understanding of numerical concepts and we examine the evidence provided on how humans acquire and represent numerical knowledge. In doing so, we will ignore issues concerning the real or fictional existence of numbers, and questions about what actually counts as mathematical knowledge. Consistent with an interdisciplinary approach to mathematics, the cognitive point of view endorsed in the present chapter examines how cognitive science can contribute to the understanding of mathematical reasoning without directly considering foundational issues. In a similar way, Tim Gowers in his contribution to this volume aims to address as a mathematician the issues raised by the philosophy of mathematical knowledge, avoiding metaphysical controversies.

The cognitive approach to the nature of numerical processing is mostly new to the philosophy of mathematics. Before twentieth-century neuroscience, the main source of information about the nature of numerical representations was mathematicians' introspection. By contrast, neuroscience offers data on the cognitive and neuronal aspects of the brain of the practising mathematician, and provides empirical findings about the processing of numerical information.

Central to the field of numerical cognition are the questions of the origin of numerical knowledge and how this knowledge is processed. There are two perspectives. The ontogenetic perspective concerns the investigation of how numerical processes developed in the brain of individuals: experimental and neuroimaging studies of numerical competence in children and in healthy and neurologically impaired adults provide the major evidence for this perspective. The phylogenetic perspective concerns the study of the biological origin of numerical knowledge: studies of infants and of animals and primates have greatly advanced our understanding on how numerical processes originate. Cognitive science suggests two levels of evolution: the biological evolution of our elementary, non-symbolic numerical abilities, and the cultural evolution of higher-level symbolic mathematics. In the course of biological evolution, selection has shaped our brain to ensure that our representations are well adapted to the external world. Arithmetic is one of these adaptations, and it has been suggested that humans and animals have an innate core representation of the magnitude of small numerosities. The second level of evolution is cultural. This level is specifically human and based on the development of symbols and language, which allows the core number system to extend beyond representations of small finite numerosities.

The present chapter discusses both levels of evolution, presenting evidence of the biological determination and of the domain-specificity of numerical competence. In section 1, we consider the evidence provided by studies with animals and primates, pre-verbal infants and impaired neurological patients. In section 2, we address the question of the relationship between numerical skills and linguistic competence. In section 3, we analyse the different uses of numbers and their cognitive significance. In section 4, we briefly discuss the limits of the cognitive science approach.

Throughout the chapter we will use the term *numerosity* to indicate discrete numerical quantities (e.g. the number of items in a set) and *numerical processing* to refer to the cognitive processes used to manipulate numbers. *Numerical knowledge* denotes the mental representation of numerosity and of other non-quantitative uses of numbers, e.g. in postcodes or in brand names like '7UP'.

1 Numerical cognition is innate and distinct from other cognitive skills

One of the main aims of studies on numerical cognition is to investigate the origin of numerical knowledge. Here we discuss the evidence in favour of the idea that numerical knowledge has an ontogenetic origin and that it is implemented by some specific neural circuits in the neurologically normal human brain. We will discuss data from infant and animal studies as well as evidence from neuropsychological and neuroimaging investigations.

Some cognitive scientists suggest that the human ability to understand and manipulate numerical quantities is distinct from other cognitive skills and other types of knowledge, and that a set of brain areas is involved in processing basic numerical knowledge. This human ability has sometimes been referred to as the 'number sense'[1] (Dehaene 1997) or the 'number module'[2] (Butterworth 1999). Evidence from three different areas of research suggests that numerical processing is domain-specific and biologically determined.

[1] '*Number sense* is a shorthand for our ability to quickly understand, approximate and manipulate numerical quantities. My hypothesis is that the number sense rests on cerebral circuits that have evolved specifically for the purpose of representing basic arithmetical knowledge' (Dehaene 2001: 16).

[2] '*The number module* is the innate core of our numerical abilities.... It categorizes the world in terms of numerosities, up to about 4 or 5. To get beyond 5 we need to build onto the Number Module, using the conceptual tools provided by our culture' (Butterworth 1999: 8).

Numerical competences in animals

The presence of evolutionary precursors in animals has been taken as evidence that our numerical competence is biologically determined. For example, it has been shown that chimpanzees can be trained to select Arabic numerals up to six corresponding to the number of items in a display, such as pens or spoons (e.g. Matsuzawa 1985). By systematically manipulating or omitting other factors, such as the shape or the colour of the objects used or visual and olfactory cues, researchers have shown that animals' performance in these tasks is driven by their numerical competence rather than by other factors. Other demonstrations of the existence of numerical competence in animals come from studies with birds. In a well-known experiment, ravens were rewarded when they successfully opened a box with the same number of spots on its lid as there were on a card placed in front of them, with the arrangement of spots on the lid and on the card systematically changed (Koehler 1951). Koehler showed that ravens could perform the task correctly and that they learned to distinguish up to 6 spots. Although humans' and animals' numerical processing differ in many respects, most starkly in the ability to use symbols, these exceptional findings have been taken as evidence that humans' numerical competence finds its precursor in our evolutionary history.

Numerical competences in babies

Are babies born with some numerical knowledge? If we hypothesize that the human capacity for numerical competence is innate, babies may show some minimal capacity to process quantities. Moreover, if our numerical skills are independent from other cognitive capacities, their early emergence in infants may be autonomous from other cognitive abilities involving language.

Given that infants cannot express themselves symbolically, the majority of the empirical studies on babies carried out in the last fifteen years are based mainly on habituation and violation of expectancy looking-time paradigms.[3] These studies demonstrate that infants are able to distinguish small sets on the basis of the number of individuals in them. For instance, in one study, 10- and 12-month-old infants watch as each of two opaque containers previously shown to be empty is baited with a different number of crackers for a maximum of six in each container (Feigenson et al. 2002). For example, the experimenter might put two crackers in one container and three in the other. To do so, the experimenter first puts one cracker and then another one in the first container (1+1), and subsequently puts

[3] *Habituation* refers to the way in which attention to a stimulus lessens over time (Wanga et al. 2004). Research based on habituation works on the assumption that the amount of time an infant spends looking at a (non-threatening) stimulus before apparently losing interest reflects information-processing efficiency in that infant. In an experimental set-up, habituation trials are usually repeated until the infant either satisfies a habituation criterion (e.g. a 50% or greater decrease in looking time on three consecutive trials relative to the infant's looking time on the first three trials), or completes a maximum number of habituation trials (e.g. 12 trials). The *violation of expectation* paradigm tests the infant's response to a novel event or to an event that supposedly violates the infant's expectation. The assumption here is that infants generally look longer at the novel or unusual event (e.g. Baillargeon 2000; Spelke et al. 1992).

one plus one plus one cracker in the second container (1+1+1). The order in which the containers are filled is varied, as is the container with the greatest number of crackers. The experimenter then observes which container babies choose. Results indicate that babies choose the container with the greatest number of pieces, with an upper limit of 4, therefore showing some understanding of numerical relations.

Another well-known experiment shows that even $4\frac{1}{2}$-month-old infants have some understanding of numerical quantities as they show expectations, for example, that 1 plus 1 makes 2 (Wynn 1992). In a simple and elegant experiment, a toy is moved from the visible side of a platform to its centre which is covered by a screen. With the same procedure, a second identical toy is then added. At this point the screen drops, in some cases revealing one toy, in other cases two toys. Wynn demonstrates that infants look systematically longer at the incorrect outcome '$1+1=1$' compared to the correct one '$1+1=2$' irrespectively of changes in other properties of the objects such as colour and shape. She interpreted this result as evidence that infants were indeed expecting two objects, therefore showing some basic numerical competence.

These and many other experiments investigating numerical competence in infants conducted in the last couple of decades show that babies are born with a capacity to recognize numerosities up to four and to react to changes in numerosity, and that they also have some arithmetical expectations. Cognitive scientists suggest that these abilities represent the rudimentary numerical capacities from which our more sophisticated symbolic numerical knowledge develops.

Loss of numerical competence in adults

Is it possible that an educated adult could no longer know what '3×9' or '$7+2$' mean and what their results are? Is it possible, moreover, that while these abilities are lost, others, such as language, memory and reading are still well preserved? Or, even more oddly, could someone be able to tell whether a gram is heavier than a kilo, but not whether 6 is larger or smaller than 8? Bizarre as they may seem, cases like these are routinely observed in neuropsychology.[4]

In the literature on numerical cognition, one case shows a particularly clear pattern of performance. Following a stroke affecting the left parietal areas, patient CG showed a severe impairment in processing numbers in a variety of tasks: she could not count, compare or add two numbers together, tell the time, make phone calls or catch the correct bus (Cipolotti et al. 1991). Remarkably, patient CG performed as well as matched controls in several verbal, memory, and semantic tasks, therefore excluding the possibility that her profound impairment with numbers was due to a more general cognitive impairment. Interestingly, another neuropsychological patient showed the opposite pattern of performance. Patient IH had a progressive neurological disorder referred to as 'semantic dementia' (Snowden et al. 1996; Hodges et al. 1992): he was not able to understand the

[4]Neuropsychology studies experimentally the cognitive abilities of patients with brain lesions of various origin (see Shallice 1988)

meaning and use of words and concepts, and the use of objects. For instance, he could not name objects such as a pen or a bicycle, and did not know what a pen is used for or where to find a bicycle (Cappelletti et al. 2001). Despite this profound semantic impairment, patient IH was perfectly able to understand and manipulate numbers in a variety of situations. He could count, read and write Arabic numbers and number words correctly, could accurately determine the larger of two quantities (e.g. two piles of salt) or two numbers (e.g. 3 and 4), perform simple (e.g. '6 × 9') and complex (e.g. '72 × 28') arithmetical operations, even at the late stages of his illness (Cappelletti et al. 2005). Other patients show similar patterns of performance, namely a selective preservation of numerical processing in the context of severe cognitive impairments (e.g. Crutch and Warrington 2002; Thioux et al. 1998).

Taken together, the several neuropsychological patients studied so far indicate that there is a dissociation (Shallice 1988) between numerical and non-numerical domains, that is to say, each can be selectively preserved or impaired. In turn, this evidence suggests that numerical knowledge is a functionally distinct semantic category within the semantic system, meaning that to some extent numerical processing is independent from other cognitive processes.

Besides showing the functional specificity of numerical knowledge, neuropsychological investigations have led to the study of which brain areas are involved in numerical processing. Most of the patients reported with numerical impairments often presented with lesions affecting the parietal lobes (e.g. Cipolotti et al. 1991; Dehaene and Cohen 1995; Lemer et al. 2003; Polk et al. 2001). On the other hand, patients with a selective preservation of numerical knowledge showed selective intactness of the parietal lobes (e.g. Cappelletti et al. 2001; Crutch and Warrington 2002; Thioux et al. 1998). This suggests that the parietal regions are involved in numerical processing. This observation has now been corroborated by other neurophysiological techniques, in particular by functional magnetic resonance imaging (fMRI).[5] This method allows us to identify the brain areas that are engaged in a cognitive task. For example, in the case of numerical abilities, it has been shown that comparing two numbers or two quantities or performing an arithmetical operation results in the activation, among other areas, of the parietal regions bilaterally and more specifically of an area called the intra parietal sulcus (IPS) (for a review, see Dehaene et al. 2003). However, the observation that the parietal areas are involved in numerical tasks does not mean that these areas are dedicated to processing numerical information: there is extensive evidence that the parietal lobes are also engaged in verbal, spatial and attention functions, among others (e.g. Dehaene et al. 2003; Simon et al. 2002).

Neuropsychological and neuroimaging studies thus indicate that number processing can be selectively preserved or impaired after a brain lesion, and suggest that some regions of the human brain—the parietal lobes—are more involved than

[5]Functional magnetic resonance imaging (fMRI) is used to study brain functions by visualizing changes in the flow of fluids that occur over time spans of seconds to minutes.

other brain areas when we use numbers. Yet some issues regarding the relation between numerical competence and other cognitive skills remain open. In particular, the relation between numbers and language has very much been the focus of debate among scientists. For instance, we may wonder whether our ability to understand and manipulate numerical quantities depend on our ability to use language. Or, as Susan Hespos puts it, 'If your language did not have words for numbers, would you be able to think about numeric quantities?' (Hespos 2004).

2 *Numbers and language*

The influence of language on our numerical capacities is obvious in many respects. Examples include the knowledge of the vocabulary of counting words ('one', 'two', 'three', ...); the use of syntax and morphology in manipulating numbers (e.g. writing 'one thousand' as '1000'); and the verbal format in which arithmetical facts and in particular multiplication problems (e.g. '4 × 5') are held, as both experimental psychology and neuropsychology have suggested (e.g. Dehaene et al. 2003). Despite these clear links, recent data have challenged the view that numerical skills depend on linguistic competence. Moreover, even studies suggesting that language has some effects on our numerical competence do not necessarily imply that there is a causal relationship between number skills and language.

Among the findings suggesting the independence of language and numbers are those coming from studies looking at their functional relationship in the brain. Since Salomon Henschen's case series in the 1920s (Henschen 1920), it has been known that impairments in language and calculation skills can occur independently. This has been confirmed by more recent studies of previously numerate adults with severe language impairments following brain lesion. Despite these lesions, these subjects show remarkably preserved numerical skills (Cappelletti et al. 2001; Crutch and Warrington 2002; Remond-Besuchet et al. 1999; Rossor et al. 1995; Thioux et al. 1998). Although there are some brain areas in common between numerical and linguistic processes (e.g. Cappelletti et al. 2007), and secondary activations of the language areas are triggered by some numerical tasks (e.g. when reading number words), neuroimaging studies show that the brain systems involved in numerical processing, mainly located in the parietal regions, are distant from the classical language areas, mainly located in the left temporofrontal regions (e.g. Dehaene et al. 2003).

Even if the link between numbers and language is not directly reflected in adult neuroanatomy, a relationship between them might nevertheless be a requirement for the development of numerical competence. Different hypotheses have been put forward about this relationship. One view holds that language shapes thoughts, and therefore also numerical understanding; this implies that speakers of different languages are guided by the grammar of their language to organize experiences differently (linguistic relativity, e.g. Levinson 1996). This hypothesis predicts that these differences should influence the way one categorizes the world, and that speakers of different languages should also think about the world differently (for a discussion of this point, see Heider and Oliver 1972; Boroditsky 2001). A stronger version of this view, the Whorfian hypothesis, holds that language determines our thoughts (linguistic determinism, Carroll 1956). From the Whorfian hypothesis it follows that animals and humans who lack language would not be able to process numerical quantities.

Recent experimental studies with illiterate populations provide important evidence on the role of language in developing numerical understanding and thereby challenge the Whorfian hypothesis. Peter Gordon studied the linguistic and numerical skills of a population of hunter-gatherers called Pirahã (Gordon 2004). In their language, Pirahã have unambiguous words only for the numbers one and two, and use words for three and four more loosely. Similarly, another population, the Munduruku Amazonian Indians, use the count words for numbers one, two, and three consistently, and for numbers four and five inconsistently[6] (Pica et al. 2004). Pirahã and Munduruku performance in a variety of numerical tasks suggested the existence of an approximate non-verbal representation of numbers similar to that of subjects with a fluent, well-developed verbal counting system. For example, they are able to compare and add (approximately) large numbers that are beyond their naming range, although they fail in exact arithmetic with numbers larger than 4 or 5 (Gordon 2004; Pica et al. 2004).

The main sources of evidence suggesting that the performance of these populations is roughly comparable to that of numerate subjects are the distance and the size effects. It is well known in numerical cognition that judging the larger of two numbers or two sets of objects is faster and more accurate the further apart the numbers or sizes of the sets are. For example, it takes less time to decide which is larger between '3' and '7' than between '3' and '2', whether symbolic (in literate subjects) or non-symbolic stimuli are used (Moyer and Landauer 1967). This is known as the 'distance effect'. In addition, the time it takes to indicate the larger of two numbers or of two sets depends also on the size of the numbers or sets. For example, it is faster to decide the larger between '1' and '2' compared to '8' and '9' even if the numerical distance between the two pairs is the same. This is referred to as the 'size effect'. These effects in the Munduruku and the Pirahã populations were similar to those shown in Western controls, suggesting that the cognitive processes that underlie their (non-symbolic) numerical manipulations are equivalent.

In conclusion, these studies suggest that populations with extremely limited or no verbal counting words have the same non-symbolic representations of small numbers that populations with well-developed verbal number systems have. Remarkably, their performance seems similar in terms of outcome (accuracy in performing number tasks) and of cognitive processes involved, as revealed by the normal distance effect in quantity comparisons. This evidence therefore allows

[6]That means that for quantities above three, they use four or five in a random fashion.

us to suggest that literate humans share with illiterate humans (and possibly with non-verbal animals) a language-independent representation of numerical quantities. These results consequently do not seem to support the Whorfian view that numerical understanding is dependent on natural language for its development.

These and other studies investigating the link between language and numerical processing have led to the observation that there seem to be two systems for representing numbers. The first is language-based, and is used to store tables of exact arithmetic knowledge (like multiplication tables) and to express exact quantities. The second is independent of language, and is used to approximate and manipulate number magnitudes. Some scientists have also suggested that these cognitive systems are implemented by two distinct brain circuits (Dehaene 1997; Dehaene et al. 1999). This hypothesis has recently been supported by a series of experiments involving bilingual college students (Spelke and Tsivkin 2001). Volunteers were divided into two groups, one schooled in Russian and the other in English, and were asked to solve a series of arithmetic problems. One of these involved exact calculations (e.g. 'Does 53 + 68 equal 121 or 127?'), and another involved approximating answers (e.g. 'Is 53 + 68 closer to 120 or 150?'). Results showed that with approximating answers, both groups performed similarly whether they were tested in English or Russian. However, a different pattern of performance emerged for exact calculations: when volunteers taught in Russian were tested in English or vice versa, they needed much longer to solve the problem (Spelke and Tsivkin 2001). This evidence strengthens the idea that the exact and the approximate systems combined are responsible for the human capacity to manipulate numbers. In general, these data provide important information on the functional architecture of numerical knowledge. They suggest that this knowledge can be selectively spared or impaired, and they also provide information about its neuronal correlates (e.g. Dehaene et al. 2003; Nieder 2005).

3 *Different uses of the same numeral*

The same numeral can have different uses. For example, the *cardinal* use of the numeral '7' may denote how many items we bought in a shop; the *ordinal* use may indicate the seventh person classified in a competition (which is not bigger than the sixth classified, but rather is after the sixth); and the *nominal* use may refer to brands or names of objects, such as the drink '7UP' (Butterworth 1999; Nieder 2005). We obviously learn how to keep these uses separate and to deal with them appropriately (Fuson 1992). However, it is interesting to explore whether some of these numerical meanings may become unavailable, for instance following brain injury.

Once more, neuropsychological patients provide important evidence in this respect. For example, one patient showed a selective (although transient) incapacity to perform number comparison tasks (e.g. to decide which is the larger of '4' and '7') despite being able to count and to tell what number comes before or after another (Delazer and Butterworth 1997). His pattern of performance therefore suggests a clear distinction between some cardinal and ordinal uses of numerals.

A more recent case showed a novel dissociation between the cardinal and ordinal use of numerals on the one hand, and their nominal use on the other. Patient JR was fast and accurate at comparing numbers or the size of objects, at counting, and at performing simple arithmetical operations, among other numerical abilities tested. Nevertheless, he could not use the same numerals when they indicated a date, a postcode or a house number, a commercial brand, or even the title of a movie. For instance, patient JR no longer knew what date Christmas day was (although he did know that '20+5' is 25); he could not state his house number (but could state the names of the road and the town he lived in); and he could correctly describe pictures depicting objects indicated by numbers such as 'Levis 501' or '7UP', but he could not name the objects themselves (Cappelletti et al. forthcoming). These cases therefore show that the cardinal ('four students'), ordinal ('the fourth student'), and nominal ('Renault 4') uses of numbers can dissociate, suggesting that they represent distinct components in the architecture of numerical knowledge.

4 *Limits of cognitive science*

We have addressed the issue of whether numerical knowledge represents a domain-specific, innate semantic category and we supported this view within the framework of cognitive science by presenting evidence from experimental, developmental, and animal psychology, from neuropsychology and neuroimaging studies. Cognitive science research nevertheless has some limitations.

One of these limitations is structural. For example, investigating cognitive processes in pre-verbal or non-linguistic subjects leaves some uncertainty regarding the nature of these processes, as they cannot be fully explored in the absence of language. Moreover, neuroimaging studies are constrained by the fact that only a limited sample of the population can be involved in imaging experiments (e.g. it is problematic to test children or animals or adults with some types of pathology). The ethical and moral limitations of studying neurological patients are even more obvious and imply that collecting large amounts of data and extensive testing sessions are not possible.

In addition to this, it is not uncommon in cognitive science to use different methodologies: different experimental tasks or stimuli, different samples or procedures are often not reproducible; and even if they are, they may result in different outcomes. These structural and methodological constraints entail that some of the information obtained about cognitive processes is neither exhaustive nor conclusive. This in turn imposes some limitations on the interpretation of the data and its generalization to different populations.

Despite these restrictions, cognitive science is constantly improving, for instance by using more systematic and quantitative methods—e.g. by collecting reaction times and by using larger samples of control subjects—and by providing more detailed information on neurological patients' brain lesions. In view of these constant improvements and its already discussed advantages, cognitive science is thus an indispensable tool to study how humans acquire and represent knowledge, and how they manipulate these representations.

5 *Concluding remarks*

In this chapter, we have presented studies investigating the cognitive basis of numerical knowledge. In particular, we discussed evidence of the origin of numerical knowledge and its cognitive and neuronal correlates. We have suggested how questions about mathematics can be transformed, as Hersh suggests, 'from philosophical questions, free for speculation, into scientific problems' (Hersh 2005: vii). Of course, if mathematics is thus seen as a human activity and as part of the real world, then mathematical knowledge becomes, like other kinds of human knowledge, fallible and subject to the constraints of evolution. In this light, cognitive science may complement and combine with other disciplines to offer a comprehensive perspective on numerical knowledge.[7]

[7]This work was supported by grants from the Wellcome Trust (to Marinella Cappelletti). We would like to thank the referees and editors for their helpful comments on this chapter.

What's there to know? A fictionalist account of mathematical knowledge

MARY LENG

According to fictionalism in the philosophy of mathematics, we have no reason to believe that there are any mathematical objects, and so we have no reason to believe any mathematical statements whose truth would require the existence of such objects. Thus, fictionalists have a quick and easy response to the problem of accounting for mathematical knowledge. While they agree that mathematical knowledge (as knowledge of acausal, mind- and language-independent objects) would be difficult to account for, fictionalists do not need to provide an account of such knowledge. If mathematical knowledge is knowledge of the properties such objects have, then, according to fictionalists, we have no reason to believe that we have any such knowledge. The fictionalist thus appears to avoid Benacerraf's challenge of providing an account of mathematical truth that is consistent with our having knowledge of such truths, since the basic assumption behind this challenge—that there are truths about mathematical objects about which we can have knowledge—is rejected.

The fictionalist is not entirely off the hook, however, when it comes to accounting for mathematical knowledge. Fictionalists claim that mathematical theories need not be true to be good, and that *consistency*, rather than truth, is an essential mark of a good mathematical theory. Furthermore, although fictionalists deny that mathematical proof from axioms establishes mathematical truth, since they deny that we have knowledge of the truth of our basic axioms, they do not deny that mathematical proof can be used to establish the *consequences* of our mathematical axioms. In fact, the notion of logical consequence is essential to standard fictionalist accounts of the applicability of mathematics to non-mathematical theories, according to which we can trust our mathematically stated scientific theories because we have reason to believe that they are correct in their non-mathematical consequences.

For fictionalists, then, it is important to account for our ability to have some logical knowledge concerning our mathematical theories. We need to account for our knowledge of the consistency of those theories, and for our knowledge of what follows logically from their axioms. But it has been objected against the fictionalist that any account of *logical* knowledge of this sort will also require us to have knowledge of mathematical objects, in one of two ways. Either consistency (and, therefore, logical consequence) is simply *defined* in terms of the existence of mathematical objects of a certain sort, so that knowledge of consistency just is knowledge of mathematical objects. Or alternatively, if we grant that the consistency of a theory does not itself require the existence of any mathematical objects, the fact remains that we often appeal to facts about mathematical objects in order to *justify* our claim that a theory is consistent. In this case, it is argued, in order to know the consistency of a theory we first need to have knowledge of some mathematical objects, and so the fictionalist has not managed to escape the problem of accounting for our ability to have mathematical knowledge after all.

Hartry Field (1984; 1991) has paved the way to a response to both of these issues, based on an understanding of modal notions such as consistency as unreduced logical primitives. By resisting the reduction of modal questions (concerning the logical possibility of a collection of axioms) to non-modal questions (concerning the existence of a model of those axioms), he avoids the claim that logical knowledge *just is* mathematical knowledge. He then argues that our successful *use* of mathematics to discover facts about consistency can be accounted for without assuming the truth of the mathematics used. If Field is right, then we can have justified, true beliefs concerning the *consistency* of our mathematical theories, even if the purported nature of their objects (as acausal, mind- and language-independent abstracta) prevents us from justifiably believing their truth.

As a fellow fictionalist, I would like to accept Field's account of logical knowledge as providing a good response to the challenges that the existence of such knowledge appear to present. My aim in this chapter is thus twofold. First, I wish to present in more detail the challenges to fictionalism that arise in the context of justifying our claims to logical knowledge, explaining why it is that the fictionalist needs to defend the claim that we can have some such knowledge, and showing how Field's account of logical knowledge responds to these challenges. In putting forward Field's non-reductive account of consistency, I will present, alongside some considerations that Field has himself used in defence of accepting consistency as a primitive modal notion, a further reason to resist the reduction of modal notions such as consistency to non-modal, mathematical correlates, arguing that, even on the assumption that there are mathematical objects, our only reason for believing that the mathematical definitions of consistency are even extensionally adequate depends on our having a prior grasp of the notion of logical possibility.

As will be shown, essential to Field's account of our use of mathematical reasoning in uncovering facts about consistency is the assumption that we can have independent reason to believe that some of our mathematical theories are themselves consistent: if the fictionalist can account for the rationality of our belief in the consistency of set theory, for example, this can account for our use of set-theoretic models to 'prove' the consistency of other mathematical theories. Field's own defence of this assumption is primarily inductive: we have some reason for believing the consistency of set theory, since we have not uncovered a contradiction yet. My second main aim will be to present some further considerations in favour of the assumption of consistency. In particular, I will argue that our psychological capacities to envisage structures satisfying our axioms provide us with a more compelling reason to believe their consistency than does mere enumerative induction.

1 *Trading ontology for modality—the fictionalist's bargain*

Benacerraf's challenge for mathematical knowledge is to account for our ability to have knowledge of acausal, mind- and language-independent mathematical objects. The fictionalist's response is to question the assumption that we do have such knowledge. What reasons, the fictionalist asks, do we have for such an assumption? The fictionalist considers arguments from pure and applied mathematical practice for the claim that we have knowledge of mathematical objects, and finds these arguments wanting. All that is needed to account for pure and applied mathematical practice is the claim that we can have some logical knowledge concerning the consequences of our theoretical hypotheses.

From the perspective of pure mathematics, we justify many of our knowledge claims on the basis of mathematical proof from axioms. However, such proofs can only provide mathematical knowledge of the truth of their conclusions if we already have reason to believe the axioms from which these conclusions are derived. The fictionalist will accept that proof can tell us the consequences our axioms, but questions the assumption that we know that those axioms are themselves true. Our interest, as pure mathematicians, in providing proofs stems, the fictionalist thinks, from our interest in discovering the consequences of our theoretical assumptions. Nothing in the practice of providing proofs from axioms requires that we have reason to believe those axioms are true. From a fictionalist perspective, then, all that needs to be accounted for in our use of proofs from axioms to justify knowledge claims in mathematics is the claim that our rules of inference will only allow us to derive a sentence P from our axioms if P does indeed follow from those axioms. And this, the fictionalist claims, is something we can know without appeal to mathematical objects.

Aside from proof from axioms, another standard justificatory technique in mathematics is the provision of counterexamples. Again, from a fictionalist perspective, provision of a counterexample does not show the falsity of the mathematical sentence in question. Rather, we look for counterexamples in order to justify the claim that a certain sentence does not follow from our mathematical assumptions. One standard example of this is the provision of a set theoretical model of our axioms in which the sentence in question does not hold. The fictionalist wishes to accept that such examples give us knowledge concerning what *doesn't* follow from our axioms. However, the claim that such knowledge requires us to have some genuine knowledge of mathematical objects (in this case, knowledge of the existence of the set-theoretic model which provides the counterexample) is questioned by the fictionalist. All we know (and all we need to know) in this case, the fictionalist claims, is that the existence of the set-theoretic model in question follows from the axioms of our set theory.

The fictionalist thus accounts for pure mathematical practice in terms of our interest in logical features of our theories: we wish to discover what follows, and what does not follow, from our mathematical assumptions, regardless of whether those assumptions are true of any objects. Furthermore, the fictionalist considers that the methods of pure mathematics do allow us to discover facts about what does and does not follow from our mathematical theories—they provide us with knowledge of logical consequence—even though she rejects the claim that these methods give us any genuine mathematical knowledge (knowledge of the mathematical objects posited by our mathematical theories). Accounting for our ability to have logical knowledge, and in particular our ability to use mathematical methods to uncover logical knowledge, is thus essential to the fictionalist's understanding of pure mathematical practice.

Turning to our use of mathematical assumptions in natural science, scientific realism (the claim that we have reason to believe that our current best scientific theories are true or approximately true) does appear to support the claim that we have some mathematical knowledge (or, at the very least, that we have reason to believe in the truth of some claims concerning mathematical objects). The scientific theories that we currently make use of include claims whose truth would require the existence of mathematical objects. If we have reason to believe that those theories are true, then we have reason to believe in the mathematical objects they posit. In fact, there are those who, following Quine, think that it is only due to their presence as posits in confirmed scientific theories that we have reason to believe in the existence of mathematical objects. Despite appearances to the contrary, mathematical knowledge is on this account empirical knowledge like any other.[1]

Against this, there are two potential fictionalist responses. The first is Hartry Field's (1980). Field is a scientific realist: he accepts that we have reason to believe our best scientific theories. Nevertheless, although he accepts that the theories we make use of from day to day in doing science do posit the existence of mathematical objects, Field claims that these are not our *best* theories. Rather, Field argues that the mathematized versions of our scientific theories are merely convenient alternatives to better non-mathematical versions, which express our actual theoretical commitments. That is, Field holds that our best scientific theories, when correctly formulated, do *not* posit the existence of mathematical objects: mathematical posits, he claims, can be dispensed with in formulating the assumptions of those theories. When the chips are down, Field hypothesizes, we should be able to state the assumptions of our empirical theories without appeal to mathematical objects.

According to Field, the fact that we can make practical use of mathematized versions of our scientific theories in the context of our day-to-day scientific practice does not count as confirmation of the truth of those theories. Rather, this

[1]See Mark Colyvan's chapter in this volume for a defence of the empirical nature of mathematical knowledge.

use is accounted for by arguing that those theories are *conservative extensions* of the non-mathematical theories that we do believe—that is, they have no non-mathematical consequences that are not already consequences of their non-mathematical counterparts. If we have reason to believe the non-mathematical versions of our empirical theories, and also have reason to believe that the mathematized versions are conservative extensions of these theories, then we will have reason to believe any non-mathematical consequence of our mathematized scientific theories, even if we have no reason to believe their mathematical assumptions. So the successful use of the mathematically stated versions of our scientific theories does not provide any confirmation of the truth of such theories, since their success is just what we should expect if we believed the non-mathematical counterparts that they conservatively extend.

Our ability to have some logical knowledge is thus central to Field's fictionalist account of the use of mathematics in applications. Field's justification of our use of mathematically stated scientific theories (in terms of their conservativeness) depends on a claim about the *consequences* of those theories: we are justified in using those theories in drawing non-mathematical consequences so long as we have reason to believe that those non-mathematical consequences are already consequences of the non-mathematical scientific theories that we take to be true. The claim that we can have knowledge concerning the consequences of mathematical assumptions is thus just as essential for this account of applied mathematical practice as it was in accounting for pure mathematics from a fictionalist perspective.

Field's project of dispensing with mathematics in the formulation of our scientific theories requires a piecemeal approach—for each mathematically stated scientific theory that we use, such a fictionalist would have to show that there is an attractive alternative theory, which does not posit the existence of mathematical objects, and which captures what it is that we really believe when we make use of the mathematized version. But there is widespread scepticism, even amongst fictionalists, that such alternatives can always be found. Mathematics provides us with very rich forms of expression: by speaking of *sets* of non-mathematical objects and the relations between various such sets, we can state in simple terms facts that are seemingly about the non-mathematical objects which would be difficult to state in non-mathematical terms.[2] Should we really expect that everything we wish to say about non-mathematical objects with the help of the hypothesis of a rich mathematical universe *can* always be said in an attractive way without making use of this hypothesis?

An alternative fictionalist approach accepts the indispensability of mathematics in formulating our scientific theories, but tackles the scientific realism that stands behind the move from our use of mathematized scientific theories to belief in the existence of mathematical objects. Such an approach challenges scientific

[2]This is most clearly the case if we restrict ourselves to first-order languages. In this case, seemingly non-mathematical assertions such as the claim that 'There are only finitely many *Fs*' cannot be expressed non-mathematically without supplementing our usual first-order resources (e.g. by allowing infinite disjunctions, or by introducing a finiteness quantifier).

realists to show that scientific evidence gives us reason to choose their hypothesis (that our scientific theories are true) over an alternative instrumentalist hypothesis (that our scientific theories are 'nominalistically adequate'—roughly, that they are correct in their non-mathematical consequences).[3] I will not consider the merits of this position here,[4] but simply note that this account of applications of mathematics also requires an account of the notion of logical consequence and our ability to have knowledge concerning the applicability of this notion.

In all of these cases, then, the fictionalist rejects the claim that we have knowledge of mathematical objects in favour of the claim that we have knowledge of the consequences of our mathematical theories. Ontology is thus traded for modality: we need not claim knowledge of mathematical objects, in order to account for pure and applied mathematical practice, but we do claim knowledge concerning the logical possibility (i.e. consistency) of collections of mathematical assumptions (with knowledge that P is a *consequence* of our mathematical assumptions understood as knowledge of the *inconsistency* of those assumptions with the negation of P).

It has been suggested, however, that this trade-off does not so much solve the epistemological problems of mathematical platonism as shift them, so that they reappear again once one considers the appeal to logical notions that the fictionalist must make. Thus, Stewart Shapiro complains that 'the epistemological problems with the antirealist programmes are just as serious and troublesome as those of realism in ontology. Moreover, the problems are, in a sense, *equivalent* to those of realism' (1997: 218). The question I want to consider is whether Shapiro is right that the fictionalist's appeal to logical knowledge itself introduces epistemological problems. Can we account for our ability to have knowledge of the consequences of our axiomatic theories, or the consistency of those theories, without reintroducing the difficult problem of our ability to have knowledge of mathematical objects? Since logical consequence and logical consistency are interdefinable (an axiomatic theory is consistent if there is no sentence P such that P and $\neg P$ are both consequences of the axioms of the theory; P is a consequence of the axioms of a theory T if the axioms together with $\neg P$ are inconsistent), I will focus on the fictionalist's claim to have knowledge of the consistency of our mathematical theories. Is this claim open to epistemological challenges equivalent to those levelled at the realist's claim that we have knowledge that such theories are true?

[3]The notion of 'nominalistic adequacy' is intended to invoke Bas van Fraassen's related notion of empirical adequacy. It is worth noting, though, that as van Fraassen defines the empirical adequacy of a theory, this is a mathematical notion (understood in terms of the relation between the mathematical models of the theory in question and empirical 'appearances'). A fictionalist clearly cannot make use of van Fraassen's mathematical resources in defining nominalistic adequacy (indeed, as Gideon Rosen (1994) has argued, it is doubtful that such mathematical resources are even available to van Fraassen, given his constructive empiricist commitment to remain agnostic concerning matters unobservable). Hence our alternative characterization of nominalistic adequacy in terms of the consequences of our mathematized assumptions.

[4]A version of this account is defended by Mark Balaguer (1998). In (2005*b*) I argued that this kind of instrumentalism is on a stronger footing than van Fraassen's constructive empiricist position, on which it is based.

2 Fictionalism and nominalism

One problem for the fictionalist arises simply by considering the *form* of the fictionalist's claim that a given mathematical theory is consistent. This would appear to be the claim that a particular object (a theory) has a property (consistency). But it looks like *theories* should themselves be thought of as abstract objects of some sort: collections of *propositions*, perhaps, or even (as proponents of the semantic view of scientific theories have claimed), collections of their *models*. If this is right, then to claim that a theory is consistent is to ascribe a property to an abstract object.

But many (probably most) fictionalists concerning mathematical theories are motivated in their fictionalism by a more general nominalism—a scepticism about the existence of abstracta of any sort. Such fictionalists consider the main problem with mathematical platonism to be just one example of a general epistemological problem with abstracta: it is mysterious how we could come to know about any such things. So, if the claim that a theory is consistent amounts to the ascription of a property (consistency) to an abstract object (a theory), then there is no advantage for the fictionalist in trading in a belief in the truth of our mathematical theories to a belief in their consistency. Fictionalism fails to escape the problem of accounting for our knowledge of some abstracta.[5]

Can the fictionalist assert the consistency of a mathematical theory without committing to the existence of the very abstracta whose existence she doubts? What can one who doesn't believe in *theories* make of the claim that a given theory is consistent? Field's answer to this problem is to understand the claim that a theory is consistent in analogy with the deflationist's understanding of the claim that a theory is true. If this analogy can be made to work, then consistency is no more to be understood a property of theories as truth is (for the deflationist). The claim that a theory is consistent will be deflated to a ground level claim in which theories and their properties play no part. Let's see how the analogy goes.

At the object level, we can express the content of a theory simply by asserting its axioms. But if we say *of* those axioms that they are *true* (or *consistent*), we semantically ascend to the metalevel, where we no longer use those axioms, but mention them, asserting of them that they have the property of truth (or consistency). It is once we make this move that the awkward question arises: what status should be given to these axioms, whose truth (consistency) we are asserting? Considering them as nominalistically acceptable sentence tokens cannot do: even for finitely axiomatized theories where we can produce tokens of each axiom, an assertion of truth (consistency) involves us in more than just a claim about particular tokens (we could not, for example, account for the kinds of inferences we wish to make on the grounds of the claim that some axioms are true if all that was meant by this was some claim about some particular tokens of those axioms). Neither do we get much solace if we consider ourselves as talking about

[5] I am grateful to Gideon Rosen for pressing this objection. Rosen makes this point in print in the context of his discussion of van Fraassen's constructive empiricism, in his (1994).

the sentence-types: after all, those sentence-types only contingently express the content of the theory whose truth (consistency) we are asserting. In asserting, of a theory, that it is true (consistent), it appears that we want to say something not about the sentences used in a typical axiomatization of our theory, but rather, about whatever it is that is *expressed by* those sentences. So the claim that a given theory is true (consistent) looks very much like the claim that the *propositions* (or worse, the *set* of propositions) that serve to individuate that theory have a particular property.

The deflationist about truth seeks to avoid this problem by rejecting the semantic ascent to talk of theories and propositions. According to deflationism about truth, what we say with semantic ascent when we speak of the truth of theories can be said at the object level, without such ascent, simply by asserting those theories themselves. All that is really said, at the metalevel, when we assert *of a theory* that it is trueis just what we can say otherwise, at the object level, by asserting the axioms of that theory. Of course, our ability to make such an assertion at the object level will depend on the resources of our object language. In particular, given that we will often want to assert the truth of theories with infinitely many axioms, we will need to augment our object language with some device which allows us to make such an assertion: a device for infinite conjunction, or a substitutional quantifier, which will allow us to assert in a single sentence all of the allowed instances of an axiom schema. But if we accept such a device into our language, then talk at the metalevel of the truth of theories can be deflated in favour of object level sentences which contain no mention of theories whatsoever.

Similarly, Field's account of consistency wishes to deflate metalevel talk of the consistency of theories to object level claims which do not mention such objects. Just as the content of the claim that the axioms of a theory is true can, according to deflationists, be expressed at the object-level without semantic ascent simply by asserting the conjunction of the axioms of the theory, so, Field argues, the claim that the axioms of a theory are consistent can be expressed at the object-level simply by asserting the logical possibility of that conjunction. Of course, our ability to assert this at the object level will depend on the resources of our object language. As well as our device for infinite conjunction, we will need to augment our language with a sentential operator, which allows us to move from a sentence P to a sentence $\Diamond P$, to be read 'It is logically possible that P'. Whenever we wish to assert that a theory is consistent (i.e. logically possible), we can express this without semantic ascent by forming the conjunction of the axioms of the theory, AX_T, and asserting that $\Diamond AX_T$.[6]

With these added linguistic resources, then, we can rewrite the fictionalist's 'knowledge' claims for mathematical theories at the object level. When the fic-

[6]Reading the claim that a theory is consistent as the claim that its axioms are (logically) possibly true, the deflationist's machinery of the '$\Diamond$' operator does for possible truth what the device for infinite conjunction does for truth: it allows us to assert the content of the claim that a theory is possibly true without the semantic ascent that would make 'possible truth' appear a substantial property of abstract theories.

tionalist claims that we can know (or have reason to believe) that a given, axiomatized mathematical theory T is consistent, her claim is that we can know (or have reason to believe) the sentence

$$\Diamond AX_T.$$

When the fictionalist claims that we can know (or have reason to believe) that P is a logical consequence of T, her claim is that we can know (or have reason to believe) the sentence

$$\Box(AX_T \supset P),$$

where the '$\Box$' operator, read 'it is logically necessary that', is defined in terms of the '$\Diamond$' operator in the usual way.

There is one clear disanalogy between deflationism about truth and deflationism about consistency. Although both require us to augment our object language in order to avoid semantic ascent and talk of theories, the additional resources required by the deflationist about truth appear much less substantial than the fictionalist's '$\Diamond$' operator. The move from finitary conjunctions to infinitary conjunctions, though by no means uncontroversial, can be relatively easily motivated through our ability to understand ordinary conjunction together with idioms such as 'and so on'. It seems pretty clear that we have a grasp of finitary conjunction—and from this can get an idea of what's being asserted when we assert an infinite conjunction. But what of the '$\Diamond$' operator? Do we grasp what it means to apply '$\Diamond$' to a sentence? And how can we know when '$\Diamond$' is applicable? Deflating talk of consistency of theories to object-level talk of the logical possibility of the conjunction of their axioms is just the beginning of the story for the fictionalist's account of consistency. It shows how the claim that a theory is consistent need not be committed to theories as abstracta, but it does little more. The main work for the fictionalist (and the main work for the rest of this chapter) is to motivate the idea that we do have a grasp of the '$\Diamond$' operator, and we can in some cases know when to apply it to sentences. In particular (keeping Shapiro's epistemological worries in mind) we must show that our grasp of the applicability of this operator in no way depends on our ability to know truths about abstract mathematical objects.

In particular, the first task for the fictionalist is to argue that this operator is understood independently of the related model theoretic notion of satisfiability (according to which a set of axioms is satisfiable if it has a model) and the related proof theoretic notion of deductive consistency (according to which a set of axioms is deductively consistent relative to a proof procedure S if there is no sentence P such that both P and $\neg P$ are derivable in S from those axioms). On the assumption that there *are* such things as sets and abstract derivations, we should hope that these three notions are co-extensive in the case of first-order theories, and closely related, though not co-extensive, in the case of second-order theories. Nevertheless, the fictionalist claims, what is said when one asserts $\Diamond AX_T$ by itself implies nothing about the actual existence of set theoretic models or abstract derivations. As such, the fictionalist's first task is to show that this notion of logical possibility is intelligible even to one who rejects the existence of abstract mathematical objects. And, since even fictionalists often use model-theoretic and proof-theoretic reasoning to *justify* claims about the applicability of the '$\Diamond$' operator, the fictionalist's second task is to argue that we can trust such reasoning even if we do not believe in the objects appealed to in standard model theory and proof theory.

Before turning to these problems, though, I would like to say a few words on the question of whether a 'deflationary' reading of the logical notions is *required* by mathematical fictionalism. Certainly, if one is motivated by a general distaste for abstract objects, understanding consistency as a property of collections of propositions will not appeal. Mathematical fictionalism and deflationism about truth and consistency make for a natural combination as a response to global scepticism about abstracta. However, it is arguable that one main motivation for mathematical fictionalism, Benacerraf's knowledge problem for mathematical objects, leaves open the question of whether one should be sceptical about abstracta across the board. If this is right, then one might be able to combine scepticism about mathematical objects with a more robust understanding of truth and consistency as properties that theories—considered, for example, as collections of propositions—can have or lack.

To see this, note that although Benacerraf's problem for mathematical knowledge is often expressed as a general puzzle about our ability to gain knowledge of abstracta, the effectiveness of the puzzle does not depend only on the *abstract* nature of the objects posited by our mathematical theories, but also on their mind- and language-*independence*. Benacerraf presents us with a puzzle—given the kinds of objects that our mathematical theories purport to be about, how could we possibly have knowledge of those objects? In presenting his worry, Benacerraf focuses on the acausality of these objects in developing its force, but it is not *just* acausality that makes mathematical objects puzzling. Indeed, acausality alone, one might think, is not enough to preclude knowledge. We might think that the case of propositions, for example, shows that we *can* have some knowledge of acausal abstracta: one can know what is expressed by a sentence in one's language (it might be argued) even though this knowledge is not *caused* by the proposition expressed. So propositional content presents (for one not already convinced by deflationism) a prima facie counterexample to Benacerraf's challenge concerning knowledge of abstracta.

Whether or not this is a genuine counterexample to the claim that we cannot have knowledge of abstracta,[7] such an example does not hurt Benacerraf's challenge to the mathematical platonist. Rather, the example helps to illustrate what it is that is meant to be *particularly* problematic about our ability to gain knowledge of abstract *mathematical* objects. If we separate out the question of what's

[7]I rather doubt that it is, since I think that there are independent Quinean reasons to be sceptical about the notion of 'the proposition expressed' by a given sentence.

said by our mathematical theories from the question of whether what is said by those theories is in fact true, there is a sense in which the first question is tractable in a way that the second isn't. In a narrow sense, the content of our mathematical theories is a function of our language and the way we use that language. As such, although knowledge of such content might be considered to be knowledge of abstracta, our ability to have such knowledge is presumably explicable in terms of our grasp of our own language. On the other hand, the question of whether our mathematical theories are true (and, indeed, the question of the wide content of our theories—of whatever it is, if anything, that they are in fact talking about) is dependent on objects whose existence is independent of our minds or language. Simply entertaining mathematical axioms does not thereby make those axioms true. We cannot stipulate mathematical objects into existence: the question of whether there are abstract objects about which our mathematical axioms express truths is outside our control. For such abstracta, acausality combines with mind- and language-independence to create particular difficulties for an adequate epistemology.

The problem of accounting for our knowledge of *mathematical* abstracta thus seems to be particularly intractable, even if one thinks that some abstracta are knowable. There is, then, room for a mathematical fictionalism which is sceptical about our ability to know truths about (mind- and language-independent) mathematical objects, but allows that we can have knowledge of abstracta such as the (narrow) propositional contents of our theories. For such a fictionalist, the claim that a theory T is consistent need not be deflated to an object-level assertion that $\Diamond AX_T$. Nevertheless, to avoid commitment to mathematical objects such as sets and abstract derivations, this meta-level claim must still be distinguished from the claim that the theory is satisfiable or that it is deductively consistent.

Whether or not one restricts oneself to a deflationary understanding of consistency, the main challenge for the fictionalist is the same: to account for our understanding of consistency as independent of the mathematical analogues of this notion. My own preference is for the deflationary version of this project, but in what follows it will often be clearer to talk at the meta-level, of the consistency of a theory T. Deflationists should note that this semantic ascent can be eliminated in Field's way, in favour of assertions made at the object level, just as talk of the truth of theories, though convenient, can be eliminated in favour of assertions at the object level of the (conjunctions of the) axioms of those theories.

3 *A new Benacerraf problem?*

Our *logical* notion of consistency (or logical possibility) is closely related to two *mathematical* notions.[8] From model theory, we have the notion of satisfiability:

[8]So closely related that our use of the term 'consistency' for the *logical notion* (of logical possibility) might be thought misleading: Stewart Shapiro (1997: 95–6), for example, reserves 'consistency' for the proof-theoretic notion of deductive consistency, uses 'satisfiability' for the model-theoretic notion of truth in a model, and introduces the notion of the 'coherence' of an axiom system for the primitive

a theory T is satisfiable if it has a model (in some chosen background set theory—let's say, ZFC). From proof theory, we have the notion of deductive consistency: a theory T is deductively consistent relative to some derivation system S if there is no sentence P such that there is a derivation in S of both P and $\neg P$ from the axioms of T. We might think that one or other of these notions captures the content of what it is that we assert when we assert that a theory is consistent. The '$\Diamond$' operator is, in this case, not so innocent after all: there is no advantage to being able to assert the consistency of a theory at the object level (by asserting the logical possibility of its axioms) if, for example, all this amounts to is an assertion about the existence of a *model* of those axioms.

If the '$\Diamond$' operator is just shorthand for some claim about abstract models or derivations, then clearly a new Benacerraf problem arises concerning our knowledge of consistency. Be it about sets or abstract derivations, knowledge of consistency will turn out to be knowledge about mathematical abstracta. Field responds to this problem by rejecting the claim that either of these mathematical notions captures the content of our claim that a theory is consistent, holding instead that consistency is a logical primitive which we grasp prior to and independently of these mathematical correlates. Thus, Hartry Field holds that even mathematical platonists should accept that 'consistency is neither a proof-theoretic notion nor a model-theoretic notion' (1991: 6). We must, he thinks, accept '$\Diamond$' as a primitive modal operator, not to be analysed away in terms of models or derivations.

Field's defence of the idea that we grasp the notion of consistency prior to its mathematical correlates takes its cue from a discussion of Georg Kreisel's (1967). We will turn to this discussion shortly, but before we do so, it will be instructive to consider the relation between attempts to reduce talk of logical possibility to non-modal talk (of models or derivations) and similar attempts to reduce talk of metaphysical possibility to non-modal talk (of possible worlds). Part of the attraction of the alternative mathematical definitions of consistency is that they each appear to reduce modal questions to non-modal ones. The question of whether a theory is consistent (logically possible) comes down to two alternatives: (i) whether there is a set-theoretic model in which the assumptions of the theory are interpreted as truths; (ii) whether there is, amongst the realm of all (abstract) derivations constructed by applying derivation rules to the axioms of the theory as assumptions, any derivation which ends in a contradiction. Accepting one of these reductions as a definition of the '$\Diamond$' operator serves to simplify our conceptual resources: we explain an operator whose applicability appears to depend on the *modal* properties of sentences in terms of entirely non-modal properties of abstract objects.

David Lewis's possible worlds theory similarly aims to explicate modal idioms by getting rid of them. For Lewis, talk of *metaphysical* possibility is replaced by talk of truth in worlds. Although Lewis advances a vast ontology of worlds,

modal notion that we've been calling 'consistency' or 'logical possibility'. I prefer to stick with talk of 'consistency' or 'logical possibility' for the logical notion, and use 'deductive consistency' and 'satisfiability' for the two mathematical analogues.

he claims that this ontology is well worth the price, since, with it, 'we find the wherewithal to reduce the diversity of notions we must accept as primitive' (1986: 4). Lewis's worlds are thus intended as *non-modal* truthmakers for modal truths: accepting them, it is claimed, obviates the need to see the world as containing any irreducible modal facts. But we might wonder about the extent to which Lewis really has been able to get rid of primitive modality from our world view. For it is arguable that, in order to guarantee the extensional adequacy of Lewis's account of the worlds as a reduction of modality (even accepting Lewis's assumption that there are possible worlds other than the actual one), we are required to place modal restrictions on what worlds there are said to be. But if the worlds of Lewis's theory really *are* the truth makers for modal facts, then no prior appeal to modality can place any restrictions on what worlds there are.

Scott Shalkowski presents the difficulty as follows:

> According to modal realism, the existence of a group of objects, the possible worlds, is supposed to be the foundation for modal truths. The existence and natures of these worlds is the primitive feature of modal reality, while the necessities and possibilities are parasitic on the nature of the set of worlds. For this account to work, there can be no modal restrictions on these worlds. Possible worlds must constrain facts of modality; facts of modality must not restrict the number and nature of possible worlds. Were God creating the entire Lewisean plurality of worlds, there would be no modal restrictions on God's act of creation. Without the worlds, there are no modal truths. The states that distinguish the modal truths from the modal falsehoods would not exist. To say that God had no choice as to which or how many worlds to create is to say that there *are* modal constraints on the number and nature of possible worlds, and this is tacitly to give up the reductive features of the modal realist's program. To admit constraints on the number and nature of worlds is to contradict the reductive modal realist's hypothesis that the existence of worlds is the prior, or more basic, feature of reality and modality the posterior, or less basic, feature. So, those who hold that possibilia provide modality with its hold on reality can give no modal argument as to how many possible worlds there are or what they are like. (1994: 675–6)

If, then, the possible worlds theory is to be a genuine reductive account of modality, the theory must place no modal restrictions on what worlds there are.

And yet it is only by placing modal restrictions on what worlds there are that this reductive theory can even be extensionally adequate. For let us suppose that Lewis is right that there are worlds other than our own. Lewis claims that all that is meant by 'It is (metaphysically) possible that P' is that P is true in some world. For this to be a good analysis of our usual talk of possibility, then, at the very least, for every P for which it is uncontroversial that it is possible that P, there will have to be a world in which P is true. Thus, for example, Lewis's account is only a good one if there are indeed talking-donkey worlds. But how do we guarantee that there is such a world, on Lewis's account? Only if we assume that *there are all the worlds there possibly could be*. Without this proviso, it would seem that, in making the worlds, God could simply have chosen not to make a talking donkey world. Then, if the possible worlds really are the truthmakers for our talk about metaphysical possibility, it would turn out that, according to the theory, it is not possible that there are talking donkeys. It appears, then, that we are faced with a dilemma. If we do not invoke modal restrictions in our theory of the worlds, in stating which worlds there are, then for all we can tell, the possible worlds theory as an account of modality might fail to respect the truth of common-or-garden modal claims. Alternatively, if we include in our theory modal restrictions on which worlds there are (e.g. stating that there are all the possible worlds and no impossible ones), we obtain an extensionally adequate theory at the expense of any genuine reduction of modal to non-modal facts. As Shalkowski sums up, 'modal realism, insofar as it is not arbitrary, is circular as a reduction of modality' (1994: 680).

Similar considerations speak against attempts to reduce assertions using the logical possibility operator to non-modal assertions about models or derivations. The clearest analogue is with attempts to reduce logical possibility to satisfiability in a mathematical model. Indeed, we can think of set-theoretic models, on this view, as the 'logically possible worlds' to which talk of the logical possibility of sentences is to be reduced. Again, in order for there to be even a chance of such a reduction being extensionally adequate, we must place some modal restrictions on what sets there are said to be. For example, since it is uncontroversial that it is logically possible that there are n objects, for any finite n, we should require that the set-theoretic universe has sets with n members, for each n. Furthermore, since it is logically impossible for there to be a set of all those sets which do not contain themselves, then we should rule this set out of our set theoretic universe: we do not want to rule in to our set theory sets which would make logical impossibilities possible. In short, in developing our theory of what sets there are (as expressed in the axioms we have chosen for ZFC set theory), we impose modal conditions: we want to rule in as many sets as there logically can be, and rule out those that are logically impossible. But if set theory provides the ground for all talk of logical possibility, such restrictions do not make sense. What is logically possible will, if model-theoretic reductionism is right, simply be determined by what sets there are. If God had taken against 7-membered collections, and decided to stop the formation of any 7-membered sets, then any sentence whose truth would require a 7-membered domain would simply be logically impossible, no matter what any of us happen to think. Again, then, model-theoretic reductionism faces a dilemma. Either we refuse to recognize any prior modal restrictions on what sets there could be, in which case we have no reason to believe the theory to respect our usual common-or-garden judgements about logical possibility. Or, alternatively, we allow that modal facts do place restrictions on the extent of the set-theoretic universe, in which case the account fails as a reduction of modal to non-modal facts.[9]

[9]Similar considerations speak against the claim that the proof-theoretic definition of consistency provides a genuine reduction of modal to non-modal talk. Clearly, the proof-theoretic definition must posit an abstract realm of derivations in order to stand a chance of working as an adequate reduction. Concrete derivations will not do, since we will not have concrete tokens of every derivation that can be constructed according to the derivation rules. The proof-theoretic account of consistency can only work if we consider it as quantifying over *all* logically possible derivations—so we will need to invoke a realm which contains all the derivations there could be.

Setting aside this general worry about attempts to provide non-modal reductions of modal talk, the *locus classicus* for anti-reductionism concerning the logical modalities such as consistency, validity, and logical consequence, and the inspiration for Field's own defence of anti-reductionism, is Kreisel (1967). Kreisel argues that the formal mathematical definitions of *validity* of a formula α, from proof-theory (α is deductively valid if α is derivable in our derivation system) and model theory (α is semantically valid if it is true in all models), should be seen as attempts to provide formal notions that are *coextensive with* a pre-theoretic notion of logical validity that we already have. The requirement placed on these definitions—of coextensiveness with our pre-theoretic notion—suggests to Kreisel that we have a workable understanding of this *logical* notion that is prior to, and independent of, either of these formal correlates. Following Kreisel, Field argues that we have a pre-theoretic notion of logical consistency, which provides an adequacy criterion for any attempt to find a formal correlate of this notion.[10]

It should be noted that Kreisel's defence of logical notions such as validity, consequence and consistency as ultimately neither model-theoretic nor proof-theoretic is presented from the perspective of a non-sceptical view of mathematical truth and knowledge. Kreisel is not a fictionalist about mathematical assertions, and is willing to take model-theoretic and proof-theoretic results at face value. So for him if model theory implies the existence of a model of a given theory, that theory does indeed have a model. If proof theory implies the existence of a derivation of a contradiction from the axioms of a theory, then, according to Kreisel, there is such a derivation. No ontological scruples prevent Kreisel from accepting these mathematical results as genuine mathematical truths, although fictionalists like Field do question them.

But even though Kreisel thinks that mathematical proofs within model theory and proof theory give us reasons for believing claims about validity and consistency in the model-theoretic and proof-theoretic senses, he thinks that there remains a question whether these formally defined properties of formal systems correspond to our *intuitive* grasp of the pre-formal notions of logical validity and possibility. Our use of model-theoretic and proof-theoretic arguments to draw conclusions about validity as a *logical* notion (rather than as a model-theoretic or proof-theoretic notion) requires justification, and such a justification will depend on showing how the formal correlates relate to our pre-formal conception of logical validity. Fortunately, Kreisel argues, there are intuitively defensible relations between our pre-formal conception of validity and its formal counterparts, which in the first-order case show that both the model-theoretic and the proof-theoretic notions are (if one accepts the truth of model theory and proof theory) both *coextensive* with our pre-formal notion. In the second-order case, there is a question about whether either notion is coextensive with the pre-formal notion. Nevertheless, even if we do not have coextensiveness, the three notions are still closely

[10]Which logical notion one takes as basic does not really matter, since consistency, validity, and consequence are all interdefinable. In what follows I present Kreisel's discussion of validity, before returning to consistency to see how the analogous discussion would go.

related in a way that explains our ability to use models and derivations to find out facts about the extension of our pre-theoretic notion of validity.

Kreisel takes our intuitive understanding of logical validity to be that of *truth in all (possible) structures*, where the notion of possible structure is itself to be considered an unanalysed, primitive notion.[11] Since actuality implies possibility, a formula that is true in all possible structures will be true in all actual ones. In particular, it will be true in all set-theoretic models (on the platonist assumption that such things exist). Hence we can know that, if α is intuitively valid (i.e. valid according to our pre-formal, logical notion, in Kreisel's terminology, Valα), then α is valid on the model-theoretic definition also (in Kreisel's terminology, $V\alpha$). On the other hand, we would not accept as adequate any derivation system that allowed us to derive (from no premises) something that could be interpreted in some structure as a falsehood, so we should also expect that (for any good derivation system S), if α is derivable in S ($D_S\alpha$), then α is also valid in the intuitive sense.

So if our derivation system is intuitively sound, we have the following relationships between Val, V and D_S:

(1) $$D_S\alpha \rightarrow \text{Val}\alpha$$

(2) $$\text{Val}\alpha \rightarrow V\alpha$$

Our intuitive pre-formal notion of logical validity can thus be seen to lie somewhere between deductive validity and semantic validity. Furthermore, having derived these relations, we can account for our use of mathematical reasoning concerning models and derivations to discover modal facts concerning validity. Even if we reject the reduction of the notion of logical validity to model theory or proof theory, these relations between the modal notion and its model-theoretic and proof-theoretic correlates make clear why it is useful to use model theory and proof theory in discovering the extension of our *Val* operator. The truth of (1) accounts for our practice of inferring the validity of α from the existence of a derivation of α, while (using *modus tollens* on (2)) we can justify our inference from the existence of a model which falsifies α to the intuitive invalidity of α. Hence the platonist's use of the mathematical notions to discover facts about the *logical* notion of validity is justified in virtue of the relations which hold between the logical notion and its mathematical counterparts, even if neither of these mathematical counterparts captures the content of our claim that a sentence is valid.

There remains the question of whether either mathematical notion is *coextensive* with the logical notion, on the platonist's assumption that there are mathematical objects such as models and derivations. In the first-order case, we can in fact show (from a platonist perspective) that all three notions are coextensive. This is due to Gödel's completeness theorem, which allows us to close up the loop and

[11]Rather than quantifying over possible structures, whatever they are, we will prefer to understand the validity of a formula in terms of the inconsistency of its negation, so 'Valα' becomes '$\neg\Diamond\neg\alpha$', with the primitive modality shifted from the notion of a possible structure to the notion of the logical possibility of a formula.

trap Val between D_S and V. Thus, according to the completeness theorem, for all first-order formulae α^1, and for a suitable choice of S,

(3) $$V\alpha^1 \rightarrow D_S\alpha^1$$

Hence, for Kreisel, although our understanding of logical validity is independent from and prior to its formal model-theoretic and proof-theoretic counterparts, in the first-order case we can show that these formal counterparts are coextensive with the intuitive notion:

(4) $$\text{Val}\alpha^1 \leftrightarrow V\alpha^1$$

(5) $$\text{Val}\alpha^1 \leftrightarrow D_S\alpha^1$$

For first-order formulae, then, the platonist can use either formal 'definition' of validity as an extensionally equivalent alternative to our pre-formal concept of validity. Nevertheless, from an anti-reductionist perspective, the fact remains that we have only been able to show the material equivalence of Val with V and D_S by appealing to our pre-theoretic notion of logical validity, which we used to justify (1) and (2). The material equivalence of the three notions does nothing to show that our concept of 'validity' *just is* that of derivability or of truth in all set-theoretic models, it merely explains why, for the platonist, all three notions are interchangeable for practical purposes.

Positive considerations *against* reductionism arise when one considers second-order formulae. In the second-order case, Kreisel's 'squeezing' argument cannot be used to show that the pre-formal notion of logical validity coincides with the two formal correlates, as we do not have a complete deduction system for second-order logic. Indeed, we know that derivability falls short of semantic validity in the second-order case, so placing informal validity between derivability and semantic validity leaves open the question of its actual extension. If we assume (as seems reasonable in this case) that formal derivability falls short of our notion of validity for second-order formulae, then there remains only the question of whether the model-theoretic notion of semantic validity captures the intuitive notion of validity. Might it be the case that the *logical* notion of validity lies properly between the proof-theoretic and model-theoretic notions for second-order formulae?

The question we are asking is whether there could be a formula which, from our intuitive perspective, is not a logical truth and yet is true in all set-theoretic models. Might it be the case that the set-theoretic hierarchy is simply not rich enough to provide a falsifying model? We know that there are limitations on the extent of the set-theoretic hierarchy: there can be no set of all sets, on pain of contradiction. In the case of first-order formulae, these limitations do not matter for judgements of validity, since we can find models of the first-order set-theoretic axioms within the realm of set theory. In this case, any first-order sentence which is false when its quantifiers range over *all* sets will also be false in these models, and so will not be declared logically valid. The second-order case, though, is a different matter. Can we be *sure* that there is no second-order sentence which is falsified only when its first-order quantifiers are understood as ranging over the entire universe of sets?[12] If not, then we cannot be sure that the model-theoretic definition of validity coincides with our intuitive notion of validity as not logically possibly false. If Val lies properly between D_S and V in the case of second-order formulae, this throws further doubt on the claim that the logical notion of first-order validity *just is* one of its coextensive mathematical correlates. For, as Field complains, '*it is only by virtue of an "accident of first-order logic"*' (1991: 4) that our reduction of validity to truth in all set-theoretic models gives the intuitively desirable results even in this case.

For Kreisel, then, the intuitive notion of validity for a second-order formula might stand properly between derivability and truth in all set-theoretic models. Nevertheless, the relationship between Val, V, and D_S can still be used to discover some facts about validity in the informal sense. Although there may be some formulas that are intuitively valid but not derivable, we can at least know (from Kreisel's realist perspective) that any formula we derive is valid in the intuitive sense (we simply apply (1)). Similarly, although there may be some formulas that are not valid but happen to be true in all set-theoretic models, we can at least know, from (2), that if we find a set-theoretic model that falsifies a formula, then that formula is not intuitively valid. Hence, on Kreisel's view, although validity (and other logical notions such as consistency and consequence) is understood independently of its model-theoretic and proof-theoretic correlates, and although these correlates are not even coextensive with the logical notion in the second-order case, we can still use model theory and proof theory to discover facts about the extension of the logical notion.

Using our '$\Diamond$' notation, we can replicate Kreisel's discussion in terms of the consistency (or logical possibility) of a theory T whose axioms are expressed by the (perhaps infinite) conjunction AX_T. On the platonist assumption that there are sets and derivations, the actual existence of a set-theoretic model of the axioms provides an interpretation of the axioms in which they are true. But actual truth certainly implies possible truth, so the existence of a model of our axioms would give us reason to assert $\Diamond AX_T$. Similarly, since our intuitive notion of the logical possibility of a theory is just that that theory implies no contradiction, then, so long as we believe that our derivation system only allows us to derive genuine implications of our theory, we can infer from the consistency of our theory the non-existence of a derivation of a contradiction from its axioms.

So, *for a platonist*, there are reasons (stemming from our intuitive notion of logical possibility) to accept the following two claims:

(6) If there is a model in ZFC of the axioms of T, then $\Diamond AX_T$.

[12] If the properties that the second-order quantifier ranges over are just the sets of their extensions, then one such sentence would be $\exists F \forall x Fx$. Any set-theoretic model of this sentence will have a set as the domain of its first-order quantifiers, with the second-order quantifiers ranging over the power set of that set. It will be true in models whose domain is a set; and yet it will be false when the first-order quantifier is understood to range over *all* the sets, since there is no set of all sets.

and

(7) If $\Diamond AX_T$, then there is no derivation of a contradiction from the axioms of T.

Furthermore, in the first-order case, a squeezing argument shows that these if-thens can be replaced by if-and-only-ifs. In the second-order case, while we don't have a squeezing argument to show that either of these notions are coextensive with '$\Diamond$', we can still use either implication to help discover facts about the applicability of '$\Diamond$'. And once more, the second-order case helps us to see why even platonists should be wary of taking either the proof-theoretic or the model-theoretic notion as providing an *analysis* of our concept of consistency. Against attempts to define consistency proof-theoretically, we know of theories (such as second-order PA together with the negation of its Gödel sentence) that are not intuitively consistent (and are unsatisfiable) from which we cannot derive a contradiction. And against attempts to define consistency model-theoretically, we can at least conceive of the possibility of a second-order sentence that is only true in the entire universe of sets, but not in any set theoretic model, so that our notion of consistency might be more permissive than the notion of satisfiability in ZFC.

A platonist, then, can believe that the notion of possible truth stands properly between the model-theoretic notion of satisfiability and the proof-theoretic notion of deductive consistency. But in virtue of her belief in (6) and (7), she can justify her reasoning about the model-theoretic and proof-theoretic notions in drawing conclusions about the possible truth of theories. In particular, she can reason from the existence of a model to the consistency of the axioms, and (applying *modus tollens* to (7)) from the derivation of a contradiction to the inconsistency of the axioms.

The fictionalist also wants to hold that the notion of possible truth stands properly between the model-theoretic notion of satisfiability and the proof-theoretic notion of deductive consistency. But for her this claim is virtually trivial. Since she doesn't believe that there are any sets to provide set theoretic models, the antecedent of claim (6) is always false, so the claim itself is trivially true. Similarly, since she doesn't believe in derivations except for perhaps concrete derivation tokens, the consequent of (7) is almost always true. But this means that, although the fictionalist can accept the truth of (6) and (7), they will not be terribly useful in providing information about the applicability of the '$\Diamond$' operator. The fictionalist can rely on the provision of actual derivations of contradictions to support the claim that a theory is not consistent (so long as she believes that our derivation system preserves implication), but she cannot use (6) to defend her use of model theory in drawing conclusions about consistency.

For the fictionalist, then, appeal to a pretheoretic notion of logical possibility, considered as a modal primitive which resists reduction to proof theory or model theory, can help to avoid the revised Benacerraf problem that arises if we accept the model-theoretic or proof-theoretic definitions as capturing what we mean when we assert of a theory that it is consistent. However, a knowledge problem remains in accounting for our ability to *know* of a theory that it is consistent, since, unless we can have knowledge of the existence of set-theoretic models, we will never be able to use (6) to justify the claim that a theory is consistent. But the use of model theory to justify claims about the consistency of axiom systems is indispensable to ordinary mathematical practice. Can the fictionalist account for the use of model theory to discover facts about consistency without assuming that model theory is true?

4 A new indispensability argument?

It appears that the fictionalist is faced with a new form of the indispensability argument for the existence of mathematical objects. The fictionalist wants to claim that, in many cases, we can have knowledge of (or, at least, rationally justified beliefs about) the consistency of an axiom system. However, in many such cases, the *justification* we give for believing that an axiom system is consistent is that it has a model in the sets. But such a justification (via the inference sanctioned in (6)) is unavailable to us unless it is assumed, after all, that we can sometimes know that the antecedent of this conditional is true. But the fictionalist's evasion of Benacerraf's problem involved claiming that nothing in mathematical practice requires us to assume that we have knowledge of the existence of mathematical objects (such as set-theoretic models). If, in mathematics, we do use conditionals of the form (6) to justify claims to consistency, then it looks like the assumption that we do have some genuine mathematical knowledge (of the existence of set theoretic models) is essential to explaining our confidence in the conclusions we draw concerning the consistency of axiomatic theories.

The fictionalist's response is to question the claim that (6) is the conditional we use in drawing conclusions about consistency. It is true, in the context of mathematics, that we often assert the existence of set-theoretic models. But the assertion that a theory has a model in the sets is always justified by showing that the existence of a model follows from the axioms of our background set theory (let's say, ZFC). And, since the fictionalist does not believe that we can have reason to know the truth of the axioms of ZFC, the fictionalist claims that the only knowledge we have in this case is the logical knowledge that the existence of a model follows from the ZFC axioms, not the mathematical knowledge that there is such a model. So, while we cannot know the antecedent of the conditional (6), what we can sometimes know is the antecedent of the following conditional:

(8) If it follows from the axioms of ZFC that there is a model of the axioms of T, then $\Diamond AX_T$.

Of course, the antecedent of this conditional makes use of our primitive modal notion of logical consequence (knowing this antecedent would amount to knowing that ZFC together with the denial of the existence of a model of T is inconsistent). Nevertheless, we can sometimes have knowledge of the truth of the antecedent of this conditional via our confidence in our derivation system. So long as we are confident that our derivation rules are truth preserving, we can rely on the production of *actual* derivations to discover facts about what follows from our ZFC axioms.

If the fictionalist can defend the truth of the conditional claim (8), then she can account for the use made in mathematics of set-theoretic models to justify assertions concerning consistency. Furthermore, given that (on the natural assumption that our derivation system respects facts about logical consequence), she can account for the use made in mathematics of derivations of contradictions to justify denials of consistency, the fictionalist will be able to account for both of the main uses we make of mathematics in uncovering the logical properties of theories without assuming that the mathematics used is itself true. Hence the indispensability of mathematical reasoning in justifying claims about consistency and logical consequence will not provide a reason for fictionalists to accept that we need to assume that we have any substantial mathematical knowledge of the truth of our mathematical theories.

5 *Defending the consistency of ZFC and PA*

But what reason have we to believe (8)? The platonist's analogue, (6), was justified in terms of the relationship between truth and possible truth. If the axioms of a theory are true (in some model), then, of course, we ought to say that they are logically possibly true, hence the inference from the existence of a model of the axioms to the consistency of the axioms. The relationship between truth and possible truth cannot help in a justification of (8). We need to consider some other property of ZFC, aside from its truth, that would give us reason to infer from the fact that ZFC implies the existence of a model of T to the consistency of T.

But of course we do not need to look far for this. For, what about consistency itself? If ZFC is consistent, then, if ZFC implies the existence of a model of T, T must be consistent too. Otherwise, the inconsistency in T would (via the model) lead to an inconsistency in ZFC, contradicting the assumption that ZFC is consistent. Finding a model in the sets provides a relative consistency proof for T: we know that T is consistent if ZFC is. So our justification of (8) is as strong as our justification for believing that ZFC is consistent.

The burden of proof for a fictionalist who wishes to use (8) to uncover facts about consistency is, then, to show that we have reason to believe the consistency of ZFC. It is, however, worth noting that, in using model theory to defend claims about consistency, we frequently use much less than the full strength of ZFC to provide models. Very often, for example, we justify our belief in the consistency of a theory by providing a model of that theory in the natural numbers. Such arguments require less than confidence in the consistency of full ZFC. For we will be able to trust such inferences to the extent that we believe that PA is consistent.

How, then, does the fictionalist fare as compared with that of the platonist, when it comes to justifying the use of model theory to draw conclusions about logical consistency? The fictionalist will think that her use of model theory to draw conclusions about the consistency of particular mathematical theories is at least as justifiable as the platonist's use, since both fictionalists and platonists need to believe in the consistency of the models provided for theories in order to deduce the consistency of the theories themselves. The platonist may cite her belief in the *truth* of set theory to grounds her belief in the consistency of set theoretic models of theories, but, if Benacerraf's worries are correct, there are no grounds for the platonist's belief that set theory is true. On the other hand, if we look for direct reasons for believing in the consistency of set theory, or in the consistency of particular models that set theory provides for other mathematical theories, it is plausible that this problem is somewhat more tractable than the problem of finding reasons for believing that set theory is true.

In this final section, then, I would like to consider briefly what reasons a fictionalist might give for defending her belief in the consistency of ZFC or in PA. If the fictionalist can defend these claims, then she can account for her uses of (8) to justify claims about the consistency of other theories.

One reason that is sometimes given for believing in the consistency of theories such as ZFC and PA is inductive. So far we've not come across a derivation of a contradiction from the axioms of (for example), ZFC, despite many mathematicians working on set theory (and, indeed, looking for constructions that might produce a contradiction) for many years. But if this is the *only* reason we have to believe the consistency of ZFC, we should perhaps be worried. As Alan Baker argues in his chapter in this volume, enumerative induction by itself provides very poor reasons for believing general mathematical claims. In the case of our failure to derive a contradiction from the axioms of ZFC, we should note that we have only produced a small finite portion of the infinitely many potential proof sequences that are allowed on the basis of the axioms and rules of inference. It's even plausible that any derivation of a contradiction from the axioms would require many more steps than we'd ever be able to write out. In this case, even if we were able to draw an inductive inference from the lack of contradictions in the derivations we have so far produced to the unlikelihood of a contradiction showing up amongst any derivations of similar length, such a conclusion would give us no cause for comfort when it comes to the question of whether *any* derivation of a contradiction is permitted, given the axioms and rules of inference. Unless we have reason to believe that the derivations we are able to produce so far are a suitably representative sample of all possible derivations, this kind of enumerative induction will provide only a very weak justification for our belief in consistency.

Another inductive justification of our belief in the consistency of ZFC appeals to the applicability of mathematics. The justification is not, in this case, by means of a mere enumerative induction, but rather via an application of inference to the best explanation. Quinean platonists hold that the best explanation of the applicability of mathematics in empirical science is its truth. Fictionalists contest this, holding instead that the best explanation of the applicability of a mathematical theory requires only that that theory is consistent with the non-mathematical facts (and therefore that it is consistent *per se*). But if the best explanation of the successful application of a piece of mathematics requires the mathematical theory that we apply to be consistent, then an application of inference to the best explanation would provide an inductive justification for our belief in the consistency of that theory. A problem with this kind of reasoning is that it leaves room for a defence of the applicability of mathematics based on something even weaker than its consistency. For example, as in the previous case, if we had reason to believe that any contradiction in our mathematical theory was only derivable in a derivation too long for humans to produce, then the best explanation of the applicability of that piece of mathematics might require not that it is consistent, but only that all statements humanly derivable from the axioms of that theory are consistent with empirical facts.

Both of these inductive justifications base our belief in the consistency of our mathematical theories on the consequences derived from those theories. In the first case, we believe the theory is consistent since none of the consequences we have derived conflict with one another. In the second, we believe it to be consistent because the consequences we have derived from it in conjunction with empirical matters of fact are consistent with empirical matters of fact. It is not surprising, then, that the same worry about the 'representativeness' of derivations causes problems in each case. If we wish to provide a justification for our belief in consistency that sidesteps this worry, we will need to look elsewhere. In particular, I would like to argue that our intuitive conception of mathematical structures can provide some grounds for believing the consistency of axiomatic mathematical theories such as ZFC or PA.

The platonist sees our intuitive grasp of the iterative hierarchy for ZFC as grounding our belief in the truth of the set-theoretic axioms. Similarly, our intuitive grasp of the natural number structure is taken to ground the platonist's belief in the truth of Peano arithmetic. Fictionalists, on the other hand, think that our ability to form a *concept* of objects of a particular kind in no way guarantees the independent existence of such objects. Hence, according to fictionalists, such arguments from intuition are poor arguments for the truth of mathematical theories. However, our ability to conceptualize configurations of objects that satisfy the axioms of set theory or number theory might well provide us with some evidence that those theories are consistent. ZFC is no random collection of arbitrarily chosen axioms, but is rather grounded in a conception of the sets, the iterative conception, that appears to us to involve no contradiction. While the appearance of consistency does not always imply genuine consistency, our ability to form an apparently coherent conception of a structure satisfying the axioms of a theory should, nevertheless, help to increase our confidence in the consistency of those axioms.

In the case of Peano arithmetic, our ability to conceive of an ω-sequence seems to provide particularly good reason for believing the consistency of PA. Indeed, it is not too much of a stretch to suggest that we are familiar with *actual* structures that satisfy the Peano axioms, not just imagined ones. We are certainly familiar with many examples of initial segments of ω-sequences (in space and time), and it is not a huge stretch to imagine these sequences continuing in the same way without end, even if we do not ourselves grasp those structures in their entirety. If we think that time is even potentially infinite, then successive time intervals can provide one model of the natural number sequence in which there is no apparent contradiction. Thus, examination of our concepts of number and of set, and finding ways to envisage how objects satisfying those concepts could be arranged (e.g. as in the iterative hierarchy) can at least plausibly count as (defeasible) evidence for the consistency of these notions. At the very least, such considerations suggest that the prospects for defending consistency claims on such grounds are greater than the prospects of defending claims concerning the face-value-truth of the theories we are considering.

6 *Conclusion*

The question of our knowledge of logical notions such as consistency, consequence and validity, is a pressing one for any fictionalist, who makes use of such notions in accounting for the nature of pure mathematical practice and for the applicability of mathematics. Thus, according to fictionalism, in pure mathematics we are involved in drawing out the *consequences* of axiom systems that we believe to be *consistent*. And concerning applications of mathematics in empirical science fictionalists will wish to claim that we have reason to believe the non-mathematical consequences of our mathematically stated scientific theories even if we do not believe their mathematically stated premises. So in either case, claims to logical knowledge are essential.

So long as the logical notions are not *defined* in model-theoretic (or proof-theoretic) terms, but are instead considered as modal primitives, there is no principled *Benacerrafian* objection to our having knowledge of logical consistency, consequence, and validity. I have suggested that there are reasons even for the platonist to resist attempts to reduce these modal notions to non-modal alternatives, both because of a general worry about the possibility of finding a non-arbitrary genuinely reductive account of modality, and because there are specific reasons in the case of the logical modalities for thinking that the reductions may be extensionally inadequate in the case of second-order theories.

Having rejected non-modal reductions of the logical notions, the question remains of whether a fictionalist is justified in using proof theory and model theory to draw conclusions about the applicability of these notions. I have argued that, provided that we can have reason for believing our derivation rules to be sound, and for believing our model theory to be consistent, we can make use of proof theory and model theory to draw conclusions about the consistency (etc.) of other theories. So the main obstacle for a fictionalist's account of our knowledge of logical notions is to justify our belief in the consistency of the models provided in model theory for other theories. But the fictionalist has various strategies for defending such a belief which, while not conclusive, at least put the fictionalist's claim to knowledge of the consistency of our mathematical theories on a better footing than the platonist's claim to knowledge that those theories are true. The fictionalist can thus account for all the logical knowledge she claims to have.

Mathematical recreation versus mathematical knowledge

MARK COLYVAN

1 *Empiricism in the philosophy of mathematics*

Empiricism in the philosophy of mathematics has a chequered history. Mill defended a version of empiricism according to which the laws of arithmetic were highly general laws of nature. Mathematical truths such as $2+3=5$ were thought by Mill to be empirical in that they tell us that if we were to take two logic books, say, and three ethics books, say, we'd have five philosophy books. But Mill's somewhat naive empiricism found itself on the receiving end of a stinging attack from Frege. This attack, I might add, was considered by many to be decisive. Frege had many complaints, but the most significant was that Mill had confused applications of arithmetic with arithmetic itself.[1]

But empiricism about mathematics arose again in a more subtle form in the work of W. V. Quine. According to Quine's version of empiricism, mathematics is empirical in the sense that the truth of mathematics is confirmed by its applications in empirical science. More precisely, Quine argues that when we empirically confirm a scientific theory, we empirically confirm the whole theory, including whatever mathematics is used. Quine is not vulnerable to Frege's attack on Mill because Quine is not confusing mathematics with its applications. Rather, Quine is invoking the applications as a reason for taking the mathematics to be true.[2] Moreover, according to this Quinean picture, mathematics is taken at face value—it's about mathematical entities such as numbers, functions, sets and the like[3]—and these entities are taken to exist because of the indispensable role they play in our best scientific theories. This argument has become known as *the indispensability argument*.

I won't defend this Quinean indispensability argument here. I've done that elsewhere (Colyvan 2001). Instead, I will highlight some of the attractive features

[1]See Mill (1947) for Mill's empiricist philosophy of mathematics, Frege (1974: §23) for Frege's attack, and Dummett (1991) for a good discussion of this exchange.

[2]See Putnam (1971) and Quine (1981*a*) for articulations and defences of this view. Interestingly, this view can be traced back to Frege (1970: 187).

[3]Although Quine's Ockhamist tendencies drive him to the view that only sets are really needed, so that's all we are ultimately committed to. There are substantial issues here though. Do we reduce the natural numbers, say, to particular sets, or will any set-theoretic construction of the natural numbers do just as well? In either case (though especially the latter) can we still be said to be employing the standard semantics? It might be argued that reductions of mathematics to set theory involve some reinterpretation of mathematical language. Thanks to Mary Leng for this point.

of the kind of empiricism that emerges from it. In particular, I'll discuss how the Benacerraf epistemological problem for mathematical realism does not have any purchase on this empiricist mathematical realism. I'll then consider, in some detail, one feature of this view that has recently come under attack.

2 *An empiricist account of mathematical knowledge*

In 1973, in a now famous paper, 'Mathematical Truth', Paul Benacerraf put voice to an epistemological concern about mathematical realism that had no doubt been around for a very long time. The concern is quite simple. If mathematical entities exist but lack causal powers, it is inexplicable how we could come to know about them. Benacerraf explicitly invoked the causal theory of knowledge as a major premise in the argument but this epistemology fell out of favour not long after the publication of Benacerraf's paper.[4] But still there is something seductive about this argument. W. D. Hart puts the point thus:

> It is a crime against the intellect to try to mask the problem of naturalizing the epistemology of mathematics with philosophical razzle-dazzle. Superficial worries about the intellectual hygiene of causal theories of knowledge are irrelevant to and misleading from this problem, for the problem is not so much about causality as about the very possibility of natural knowledge of abstract objects. (Hart 1977: 125–6)

But what then is the worry about abstract objects? What is it about abstract objects that suggests that it's impossible to have knowledge about them? In my view, the most cogent post-causal-theory-of-knowledge version of this argument is due to Hartry Field. He captures the essence of the Benacerraf argument when he puts the point in terms of explaining the reliability of mathematical beliefs:

> Benacerraf's challenge—or at least, the challenge which his paper suggests to me—is to provide an account of the mechanisms that explain how our beliefs about these remote entities can so well reflect the facts about them. The idea is that *if it appears in principle impossible to explain this*, then that tends to *undermine* the belief in mathematical entities, *despite* whatever reasons we might have for believing in them (Field 1989: 26, emphasis in the original).

This challenge is usually understood to be to account for the reliability of the inference from 'mathematicians believe that *P*' (where *P* is some proposition about some mathematical object(s) to '*P*', while explicitly detailing the role that the mathematical entities in question play in this reliable process. But stated thus, we see that a substantial question is being begged against Quine and other epistemic holists. Epistemic holists hold that we do not justify beliefs one at a time. Rather, we justify packages of beliefs. How large that package is depends on how radical is the holism. Quine wavered a little on this issue, at times suggesting that it was the whole system of beliefs that was justified, at other times, he more reasonably allowed for (largish) proper subsets of our beliefs to be justified.

[4]Mark Steiner (1975) was one who took issue with the causal theory of knowledge.

So from the epistemic holist point of view, this interpretation of the Benacerraf–Field challenge is simply question begging—it assumes that mathematical beliefs are justified one at a time.

To avoid begging questions, let's take a more charitable reading of the Field version of the epistemic challenge, according to which the challenge is to explain the reliability of our systems of beliefs and to explicitly articulate the role the world plays in this reliable process. Note that we can't ask after the roles of individual objects any longer. Since we are interested in the justification of whole systems of beliefs, the best we can do is to ask after the joint roles played by collections of objects in reliable belief acquisition. In some cases this collection might be so large as to include the whole world (including, of course, whatever abstract objects there are).

Once the challenge is put this way, we see that Quine has already answered it: we justify our *system of beliefs* by testing it against *bodies of empirical evidence*. No distinction is made between mathematical beliefs and other beliefs. Our beliefs form a package that performs well against the usual standards of theory choice and that's all that matters. Any challenge to provide an account of only the mathematical beliefs is illegitimate. According to the holist, mathematical beliefs are justified in exactly the same way as other beliefs: by their role in our best scientific theories and these, in turn, are justified by appeal to the usual criteria of theory choice (empirical adequacy, simplicity, explanatory power, and so on).[5]

Still, it might be thought that all this talk of holism is beside the point. The point is that mathematical objects don't seem to contribute to the success of theories in the same way as, say, electrons, and *this* is what is in need of explanation. But again this way of stating the problem requires singling out the mathematical entities and asking after them. The thorough-going holist would deny that it makes any sense to do this, and they would thus reject the assumption underwriting the challenge. This is not dodging the issue or introducing philosophical 'razzle-dazzle'. Holism cannot be bracketed for the purposes of getting an objection up against the holist. There may be something to the epistemological challenge, but until it is formulated in a way that has some bite against the Quinean empiricist, I'm inclined to suggest that the burden lies squarely with those who believe there to be a problem here for the Quinean.

Finally, a couple of points of clarification. Quine is interested in *justification* and the Benacerraf–Field challenge is seeking an explanation of the *reliability* of the belief-forming process. Have I missed the point by focusing on Quinean holistic justification? I don't think so. In the current context, at least, we are assuming that the methods of current science are reliable methods for forming beliefs. Indeed, we are justified in believing our scientific theories, in large part because we believe that these theories were arrived at by reliable methods. Of course this assumption of reliability might be mistaken, but to push in that direction would be to mount a more general sceptical challenge, and I take it that the Benacerraf–

[5]This line of thought is advanced in Rosen (1992: ch. 3) and Colyvan (2006).

Field challenge is supposed to be a particular challenge for mathematical knowledge. Moreover, it should be noted that such general sceptical challenges typically ask us to step outside our best scientific theories and explain the reliability of (or otherwise justify) those theories from some external vantage point. This, according to the Quinean naturalist, is not possible, so any general challenge thus formulated is again question begging (but this time against naturalism). If the request for justification is in terms of intra-scientific justification, then, as Quine has argued in various places (e.g. in (1974: 3)), it is a challenge that can be met by invoking our best science (theories of optics, psychology and so on). We may be stuck in Neurath's boat, but there are some powerful tools available to us on that boat.

If we accept all this, and we admit that the Quinean epistemic holist has a good reply to the Benacerraf–Field epistemic challenge, a serious issue arises about those portions of mathematics left unapplied. After all, on the view under consideration, it's only the mathematics that finds itself indispensable to our best scientific theories that is justified. The rest, it would seem, must have a different status. I address the issue of the status of unapplied mathematics in the next section.

3 *Unapplied mathematics as mathematical recreation*

Unapplied mathematics is something of a nuisance for Quine. It can't be justified by the same means as applied mathematics, since it's precisely the applications that provide the justification. Moreover, Quine's empiricism won't allow other (non-empirical) means of justification, so it seems that unapplied mathematics does not have the same status as applied mathematics.[6] Applied mathematics is treated realistically—its propositions are believed to be true and the objects quantified over are treated as real—while unapplied mathematics, it would seem, must be (at best) treated agnostically. Charles Parsons (1983) pushes precisely this point and in reply Quine argues that it is reasonable to treat realistically a bit more than the mathematics that does in fact find itself indispensable in applications. We should include whatever mathematics is required for 'rounding out' that which is applied. The latter includes a great deal of set theory, since set theory is usually thought to underwrite most contemporary mathematics, both applied and unapplied. But how much set theory enjoys the exalted position of 'justified'? And what is the status of the rest? In response to the first question, Quine's Ockhamist sympathies come to the fore and he draws the line at the constructible sets: $V=L$. According to Quine, the demand of simplificatory rounding out of applied mathematics may be thought to extend only so far as the constructible sets. As for the second question, Quine bites the bullet Parsons offers and admits that

[6]I'm using 'applied' (and 'unapplied') here in the intuitive sense. In the mouths of mathematicians, 'applied mathematics' corresponds (roughly) to numerical methods (as opposed to pure, analytic methods).

'magnitudes in excess of such demands, e.g. $\beth_\omega$ or inaccessible numbers' should be looked upon as 'mathematical recreation and without ontological rights (1986: 400).[7]

Although I want to defend the distinction between that mathematics which we treat realistically and recreational mathematics, I will part company with Quine on a couple of issues here. First, I think that Quine makes it sound as though there are two quite different kinds of justification at work here. Lower mathematics (the lower reaches of set theory, analysis, and the like) is justified by the indispensable role it plays in our best scientific theories; the upper reaches of constructible set theory (transfinite arithmetic and so on) is justified by quite different means. The latter is justified by something akin to an act of charity: it is justified by simply being close enough to the mathematics that is applied. That is, according to Quine, accepting the upper reaches of constructible set theory is the most natural and simple way to round out the mathematics that is applied. But I think we can do better than this. I note that indispensability is transitive. If a nail gun is indispensable to building houses and building houses is indispensable to building suburbs, then a nail gun is indispensable to building suburbs. Similarly for mathematics. If transfinite set theory is indispensable for analysis and analysis is indispensable for physics, then I say transfinite set theory is indispensable for physics. Perhaps this is what Quine had in mind with his notion of 'simplificatory rounding out'. In any case, this is the justification for the higher reaches of set theory that I endorse. Understood this way, there is only one mode of justification: playing an indispensable role (either directly or indirectly) in our best scientific theories.

But is indispensability really transitive? Gideon Rosen (private communication) has questioned this claim. Rosen suggests that although large cardinals, say, might be indispensable for our best theory of real numbers, and real numbers might be indispensable for our best theories of space–time, it need not follow that large cardinals are indispensable for the physics of space–time. Physicists might

[7]Quine later refined his position on the higher reaches of set theory and other parts of mathematics not applicable to natural science:

> They are couched in the same vocabulary and grammar as applicable mathematics, so we cannot simply dismiss them as gibberish, unless by imposing an absurdly awkward gerrymandering of our grammar. Tolerating them, then, we are faced with the question of their truth or falsehood. Many of these sentences can be dealt with by the laws that hold for applicable mathematics. Cases arise, however (notably the axiom of choice and the continuum hypothesis), that are demonstrably independent of prior theory. It seems natural at this point to follow the same maxim that natural scientists habitually follow in framing new hypotheses, namely, simplicity: economy of structure and ontology (Quine 1995: 56).

And after considering the possibility of declaring such sentences meaningful but truthvalueless, he suggests:

> I see nothing for it but to make our peace with this situation. We may simply concede that every statement in our language is true or false, but recognize that in these cases the choice between truth and falsity is indifferent both to our working conceptual apparatus and to nature as reflected in observation categoricals (Quine 1995: 57).

look for different things in their theories than does the mathematician. Similarly, housing developers may look for different things in a suburb than the carpenter building the houses. While houses are indispensable to suburbs, houses built with nail guns may not be. The carpenter might be interested in strength of construction while the developer is interested in speed of construction, for example. In response, I suggest that there is an equivocation here involving the word 'best'. While it seems right that the best suburb (in the developer's sense of 'best') need not be built from the best houses (in the carpenter's sense of 'best'). But if we insist on the same sense of 'best' throughout Rosen's concern is laid to rest and transitivity is restored. The question is whether, in the scientific examples at issue, we can insist on the same sense of 'best'. Rosen seems to be on firm ground here, for surely set theorists and physicists look for quite different virtues in their best theories. Indeed, Penelope Maddy (1997) argues convincingly that set theorists do not seem to value parsimony as a virtue at all. Set theorists want as many different structures as possible. Physicists, on the other hand do seem to value parsimony.

In response, I suggest that issues concerning disciplinary expertise save the transitivity of indispensability (in the scientific context, at least). Physicists might value parsimony in their physical theories but when it comes to deciding what the best theory of the real numbers is, that's a mathematical question and it is decided by mathematical standards. Sure these standards are different from those of the physicist, but it's the mathematicians who decide what the best theory of the reals is. If the mathematicians decide that a large cardinal axiom is indispensable for this theory, then so be it. The physicists do not get to apply their standards here and they do not have the relevant expertise to do so. Now, if the best theory of space-time requires the real numbers, then whether the physicists like it or not, large cardinals are indispensable to real number theory (or so we are assuming for the point of the example) and so large cardinals are also indispensable to theories of space-time. So even if different theoretical virtues are respected in different parts of science, issues concerning disciplinary expertise ensure the transitivity of indispensability.[8]

A couple of points to note in relation to this somewhat more liberal understanding of indispensability. It may turn out that very little, if any, mathematics is unapplied on this account. After all, on this account, for a branch of mathematics to be unapplied, it must be totally isolated from the main body of mathematical theory; it must not find applications in any chain of applications that bottoms out with applications in empirical science.[9] Also, on this account, it is not so clear that one can draw the line at constructible set theory. The debates in set theory over the various large cardinal axioms, for instance, seem to be about the most natural way to extend ZFC so as to have pleasing and intuitive consequences for

[8]And if you accept Quine's view that all mathematics is set theory, then the transitivity of indispensability might be established more directly: real numbers are sets, so set theory is indispensable to space time theory.

[9]Or at least, it must be dispensable to the main body of mathematical theory. More on this shortly.

lower set theory and higher set theory alike. So even the most abstract reaches of set theory may yet turn out to be applicable, in this extended sense of applicable.[10] At the very least, the view of applications I'm endorsing here does not ignore the higher reaches of set theory (as Penelope Maddy (1992) once complained of Quine's philosophy of mathematics). Finally, I note that what is indispensable now may be dispensable tomorrow. Just as nail guns replaced hammers in building houses, we might find replacements for some of the mathematics that we now think of as indispensable.

The second way I depart from Quine on the issues under consideration follows from this. As I've already noted, it is Quine's Ockhamist sympathies that compel him to keep his ontological commitments to a minimum. While I too have such sympathies in some areas of metaphysics, it's not clear that Ockhamist considerations are appropriate here. After all, Quine is already committed to a very large infinity of abstract objects, so why baulk at a few more?[11] Of course the very nature of the Quinean argument invoked to establish the existence of mathematical entities restricts discussion to those mathematical entities that are indispensable for science. The real issue then is how much set theory is needed for science. (After all, Ockham's razor implores us not to multiply entities beyond *necessity*.) If what I suggested in the previous paragraph is correct, much more than constructible set theory is necessary, so even Ockhamists like Quine can countenance more than just the constructible sets. For the record, my position on this is to side with the majority of set theorists and accept that set theory really does need more than the constructible sets. I thus reject $V=L$. How much more? I take it that the jury is still out on that issue. But I certainly don't have in-principle objections to set theory extended by some a large cardinal axiom such as MC ('there exists a measurable cardinal').[12]

These may seem like major departures from Quine's position, but I think not. On the first issue, the way I justify the higher reaches of set theory is only superficially different from Quine's, if at all. Although Quine never emphasized the chains of applications, this may well be what he had in mind when he suggested justifying set theory up to $V=L$. On the second issue, the disagreement is more substantial. Quine is very restrictive about how much set theory we should treat realistically. I, on the other hand, am endorsing a much less restrictive attitude. But even this difference is not as significant as it might at first seem. I take it that there's nothing in the core Quinean doctrines that drives him to accept $V=L$. He needs to draw the line between applied and unapplied mathematics in a neat and

[10]Though see Feferman (1992) for a defence of the view that not very much mathematics is required for empirical science.

[11]See Burgess and Rosen (1997) for a very nice discussion of Ockham's razor in the context of mathematics.

[12]Also, I do not share Quine's view that a bivalent logic (presumably classical first-order logic) applying to all sentences of the language is simpler than some of the alternatives. (See n. 7 above, second quotation from Quine.) I will resist the temptation to take up this interesting issue here, since it's somewhat tangential to my main purpose.

convenient way (and, as I've already noted, in accord with his Ockhamist sympathies). I too have to draw the line somewhere; it's just unclear to me where that somewhere is, and I'm inclined to draw it a little further along than Quine.

Let's now return to the points on which Quine and I agree. We both accept that mathematics is justified by the indispensable role it plays in our best physical theories. We both accept that such justification does not extend to all contemporary mathematics. At least we both agree that it is conceivable that some portions of contemporary mathematics are without this kind of justification. As I've already noted, Quine takes a fairly hard line with regard to such mathematics and gives it the status of mathematical recreation. And on this too we agree. But it is important to note that in calling it 'mathematical recreation' Quine is not dismissing it. Mathematical recreation remains an important part of mathematical practice. It should not be thought of as mathematicians merely having a good time and engaging in a pastime quite distinct from their normal practice. Mathematicians engaged in mathematical recreation are much like theoretical physicists exploring different possible physical theories. Physicists, for instance, studying the Schwarzchild solution to Einstein's equation or Newtonian celestial mechanics might be thought to be engaged in 'recreational physics'. They are most certainly not studying anything real—we simply do not live in a Schwarzchild or a Newtonian universe. Nor are these physicists just having a good time and leaving behind standard practice. Investigating such non-actual solutions is an important part of standard scientific practice.

What is the point of engaging in recreational physics and recreational mathematics? There are many reasons for pursuing such activity. By studying the non-actual, we often come to a better understanding of the actual (by, for instance, coming to a better understanding of the underlying laws). We might be deliberately making simplifying assumptions because the actual situation is too complicated. We might not be sure of what is actual and so taking a pluralistic attitude means that all bases are covered, so to speak. Or it might be simply intellectual curiosity. The bottom line is that mathematical recreation, like other forms of theoretical scientific investigation, should not be thought of as second class or *mere* recreation.[13]

To sum up my position. I accept that there is a distinction between unapplied mathematics and applied mathematics—even given my very liberal sense of application via chains. I accept that applied mathematics should be treated realistically and with unapplied mathematics we have no reason to treat it this way. Unapplied mathematics is akin to theoretical investigations elsewhere in science and, as such, is an important part of mathematical practice. It is also important to note that while many branches of mathematics are at least initially pursued as recreational, they nonetheless end up being applied. Mathematics can thus change

[13]Indeed, the phrase 'mathematical recreation' is a little unfortunate. Perhaps 'theoretical' or 'speculative mathematics' would have been better, but the phrase 'mathematical recreation' is already in the literature, so I'll stick with it.

its status with regard to the recreational–non-recreational divide. While there remains something of a mystery as to how mathematics pursued by apparently a priori means and without regard to applications can end up being applied,[14] there is no doubt that this happens. On the empirical account of mathematics I'm defending here, applications make all the difference. Once a branch of mathematics finds an application, it should be treated realistically.

4 *Is all mathematics recreation?*

Mary Leng has argued that allowing some mathematics to be treated as recreation and without ontological rights, leads to a slippery slope to all mathematics being recreational. First she voices a general worry to soften us up for the argument to follow. She notes that there's nothing in mathematical practice that distinguishes between recreational mathematics and literally true mathematics. I agree. Leng then goes on to suggest that:

> Considered in this light, Colyvan's distinction between literally true mathematics and merely recreational mathematics begins to look like a distinction without a difference. The literal truth of a mathematical theory will make no difference to how a mathematician goes about working in that theory (2002: 408).

But just because there's no distinction to be found in mathematical practice, does not mean that this is 'a distinction without a difference' (Leng 2002: 408). As I've already pointed out, I take it that there's no methodological difference that cleanly marks the boundary of recreational physics from other parts of theoretical physics. Leng is right that this is not a methodological distinction, but that does not mean that it's not a distinction at all. Still her main point is correct: mathematical methodology does not recognize the recreational–non-recreational distinction. That distinction is extraneous to mathematical practice. It is determined by which parts of mathematics find indispensable applications in physical science. Applied mathematics is where the action is. On that Leng and I agree.

Now to Leng's main argument. Crucial to her argument is what Leng calls 'the modelling picture' of mathematical applications.[15] According to this picture, mathematics is never assumed to be literally true in any applications; it is judged to be adequate or inadequate for a particular application and that's the end of it. The role of the mathematics is to represent particular features of the physical system under investigation and it may do this well or poorly. According to the modelling picture, mathematics can perform this representational function irrespective of the truth of the mathematics in question. Indeed, on the modelling view, mathematics is not confirmed or disconfirmed at all. At best, the adequacy

[14]This feature of mathematical practice is often referred to as 'the unreasonable effectiveness of mathematics' (Wigner 1960).

[15]A similar account of mathematics in applications has been advanced by Christopher Pincock (2004*b*; 2004*a*) and criticized in Bueno and Colyvan (n.d.).

of the representation is confirmed. So, for example, when we find space–time is not Euclidean, we do not claim to have disconfirmed Euclidean geometry *as a mathematical theory*. Rather, we claim to have disconfirmed the adequacy of Euclidean geometry as a suitable representation of space–time, and the latter is quite different.[16]

Using this modelling picture of mathematics in applications, Leng then goes on to argue that all mathematics is recreational.

> If Colyvan is right (and I think he is) that mathematics that is not assumed by science to be true should be seen as recreational (and given some important status as such), then it follows from the modelling picture of the relationship between mathematics and science that all mathematics is recreational. (2002: 412)

This argument can be spelled out thus:

> Premise 1. Empiricism holds that mathematics with no empirical confirmation should be viewed as merely recreational.
> Premise 2. The modelling view of applications has it that when we use mathematics to represent (or model) non-mathematical phenomena, all that is confirmed is that the mathematics allows for a good representation, not that it is true.
> Premise 3. The modelling view accounts for all applications of mathematics.
> Conclusion: Therefore, all mathematics is recreational.

This is a very interesting argument. Although I will argue that it is ultimately flawed, I think Leng's argument raises important issues that cut right to the heart of the indispensability argument and the subsequent debate. As I've already indicated, I accept premise 1 (with the earlier provisos about how the more remote reaches of mathematics might be confirmed indirectly). Here I'll take issue with premises 2 and 3.

First, let's look at premise 2 and, in particular, the important role played by Sober's (1993*a*) argument against mathematics accruing confirmational support. Sober argues that the truth of mathematics is never placed on the line—if a mathematicized physical theory such as Newtonian mechanics turns out to conflict with experience, then the mathematics employed is never thought to be shown to be false. At worst, the mathematics is simply thought to be an inappropriate way to represent the theory in question. But this, Sober suggests, shows that mathematics is not really being empirically tested at all. So, in particular, there is no reason to think that mathematics employed in a successful empirical theory enjoys whatever confirmational support the theory accrues.

In Sober's paper (1993*a*), the argument is cast in terms of the contrastive empiricist theory of confirmation.[17] Sober goes on to argue that the main point

[16]Here Leng invokes Elliott Sober's (1993*a*) criticisms of the indispensability argument and the holistic picture of confirmation. More on this shortly.

[17]This theory compares likelihoods, $\Pr(E|H_1)$ and $\Pr(E|H_2)$, of two competing hypotheses H_1 and H_2 in the light of some evidence E. Contrastive empiricism suggests that H_1 receives greater confirm-

against the indispensability argument is independent of this particular theory of confirmation, and I take it that this is why Leng doesn't address the issue of the plausibility of contrastive empiricism. Be that is it may, Sober's argument is not independent of separatist confirmation theory. That is, he assumes that we can confirm or disconfirm hypotheses one at a time. But as we've already seen, this is a point that Quine denies. So Sober's objection is question begging.[18] Indeed, this can be seen from the fact that other, clearly empirical hypotheses, are never called into question when a theory confronts recalcitrant data. As Michael Resnik points out (1997: 168) various conservation laws seem immune from revision and yet it is unreasonable to deny empirical content to such principles. What's going on here is that some parts of the theory (such as mathematical principles and conservational laws) play a rather central structuring role in the scientific theories in which they appear. Sober is right that they rarely get called into question when the theory encounters recalcitrant data. But this is because, according to the holist, at least, the theory as a whole is untenable. But it is a mistake to conclude from this that every part must share the blame equally. Typically, when a theory conflicts with evidence it is only a small part of the theory that needs to be revised. There's still a substantial issue as to why it is never (or at least *almost* never) mathematics that is revised.[19] But the fact that mathematics rarely takes the fall is no reason to conclude that mathematics should not take at least some of the credit in successful theories.

Now to Premise 3. Crucial to Leng's main argument is the assumption that the *only* role mathematics plays in science is representational (hence the 'modelling picture' of mathematical applications). The central idea of this view of the mathematics–science relationship is that we have some physical system such as a population of organisms, we represent the number of organisms by a mathematical function such as the logistic function—or more commonly, we describe some features of the function in question by stating the appropriate differential equation).[20] If we then notice that the mathematics produces the wrong answers, we say that the mathematics in question was not appropriate. We do not reject the theory of differential equations, say. On this account of the relationship between mathematics and science, mathematics provides nothing more than a convenient set of representational tools. But such an account seems to seriously understate the role of mathematics in science. I've argued elsewhere (2001: ch. 3) that mathematics may contribute directly to explanations in science. If this is right, then

ational support from the evidence E if $\Pr(E|H_1) > \Pr(E|H_2)$.

[18]Of course, both Sober and Leng are attacking confirmational holism so it's also question begging for the Quinean to invoke confirmational holism as a response to them. At best they have achieved an unsatisfying stand off. I also think that there are other problems with Sober's argument, but since I've dealt with these problems elsewhere (Colyvan 2001: ch. 6), I won't revisit them here.

[19]See Resnik (1997: ch. 7) for more on why this should be so.

[20]The logistic equation, for instance, is usually represented as a first-order differential equation: $dN/dt = rN(1 - N/K)$, where N is the population abundance, t is time, r is the growth rate, and K is the carrying capacity.

mathematics is more than a *mere* representational tool and the modelling picture is wrong. After all, if mathematics is contributing directly to explanations, it is hard to see how anyone who accepts inference to the best explanation can accept the explanations yet deny the truth of the mathematics.

I now present a few examples where I take it that the mathematics in question is doing more than merely representing; it is also explaining. These examples, thus undermine the plausibility of premise 3 of Leng's argument.

Example 1. Consider the ancient problem of squaring the circle: using only compass and straight-edge, construct a square with the same area as a given circle. Here we can represent the various physical activities (marking off lengths with the compass and drawing lines with the straight-edge) mathematically. Leng is right in suggesting that the mathematics is modelling the physical activities. But she is wrong in suggesting that that's all the mathematics does. For the construction in question, as we now know, is impossible, and the explanation of why it is impossible is that π is transcendental. The mathematics, it would seem, is not only modelling but also *explaining* the impossibility of certain physical activities.

Example 2. A mountaineer sets out at 6.00 am from base camp with a load of supplies and arrives at the top camp later that same day. The following day the mountaineer returns to base camp, again leaving at 6.00 am. It turns out that there will be a point on the mountain that the mountaineer will pass at the same time on both days. Why should there be such a point on the mountain? If we represent the physical situation in the obvious mathematical way, a fixed point theorem then guarantees that there will be such a point on the mountain. Again it seems that the mathematics in question is doing more than merely modelling; it is also explaining the existence of a physical event, namely, the location of the mountaineer at the same height on the mountain at the same time on the two days in question. This case is interesting because although the fixed-point theorem guarantees that there will be some such point on the mountain, it doesn't explain why it's any particular point. An explanation of *that* fact will presumably proceed via a detailed causal story of the mountaineer's slog up and down the mountain. But this does not change the fact that the explanation of why there should be any such point is a topological explanation.

Example 3. The previous example, in particular, is a little artificial, so let me provide a real example of a phenomenon that scientists feel is in need of explanation. Evolutionary biologists are puzzled by the presence of apparently maladaptive traits, such as altruism. As Elliott Sober (1993*b*: 98–102) points out, altruistic individuals are less fit than non-altruists in a given population, so we would expect natural selection to force a decline in the relative frequency of altruism. But altruism is alive and well. How can this be? One crucial piece of the explanatory story may well be purely mathematical in nature, and involves nothing other than simple facts about inequalities, addition and division. It is common sense that if a trait is declining in relative frequency in every group, then it is declining in relative frequency overall. But for all its intuitive plausibility, this piece of reasoning is fallacious. Simpson's paradox (Malinas and Bigelow 2004) shows how a trait can be less fit relative to each of a number of groups, yet fitter relative to the ensemble of groups.[21] To take a simple example, suppose that there are two groups, A and B. In group A altruists outnumber non-altruists by 200:50. In group B there are 100 of each. After selection we find that in group A there are 150 altruists and 45 non-altruists, and in group B, there are 15 altruists and 20 non-altruists. So the fitnesses of altruists in groups A and B are 0.75 and 0.15 respectively. The fitnesses of the non-altruists are 0.9 and 0.2 respectively. As you would expect, in each group, the non-altruists are fitter. But look what happens in the combined population $A+B$. Here the fitness of the altruists is 0.55, whereas the fitness of the non-altruists is 0.43. The explanation for this peculiarity is simple and it's mathematical: although $a/b > c/d$ and $e/f > g/h$ it does not follow that $(a+e)/(b+f) > (c+g)/(d+h)$.[22]

As seductive as the modelling picture of the relationship between mathematics and science is, it ignores important aspects of this relationship.[23] To be sure, there are many cases where mathematics is used to represent and nothing more. Leng's example (2002: 411) of population dynamics may be one such case. Indeed, Ginzburg and I have suggested (2004: 31–3) that ecologists quite rightly resist mathematical explanations of ecological facts—they hold out for ecological explanation. Such examples of mathematics in science suit Leng well. But since she is offering a general account of the mathematics–science relationship, she needs to argue that in *all* applications, mathematics merely represents. In particular, she needs to give an account of cases like those above (and others I present in (2001; 2002)) where mathematics contributes to scientific explanation. Until such an account is forthcoming, we have good grounds to reject the modelling picture of the mathematics–science relationship. At least it cannot be the whole story and so premise 3 of Leng's argument should be rejected. And with that premise goes Leng's conclusion that all mathematics is recreational.

An important issue emerges from this debate though. A great deal of the early literature on the realism–anti-realism debate in the philosophy of mathematics focused on the mere fact that mathematics has applications in science. Leng is right to follow Maddy (1997) and others to look more carefully at the details of those applications. But the relationship between science and mathematics is complex and multifaceted. I don't think that the modelling picture does justice to the variety of applications and the complexity of the relationship between science and mathematics, though I'm not offering any account in its place here. I'm

[21] For present purposes, we take fitness of a group to be the ratio of the number in the group before selection to the number after selection.

[22] See Colyvan (2001) and Lyon and Colyvan (2008) for other examples of mathematical explanation and also Alan Baker's (2005) very nice example of some elementary prime number theory explaining facts about Cicada life cycles.

[23] Here I've focused on one aspect of what the modelling picture ignores: explanation. But elsewhere (2001; 2002), I've suggested that mathematics can contribute to other scientific virtues such as unification and even novel predictions. Although, see Melia (2002) for disagreement on the unification claim and Leng (2005*a*) and Leng (2008) for disagreement about the philosophical significance of mathematical explanation.

inclined to think that a great deal more work needs to be done on this issue, with detailed case studies on particular applications.[24] At this stage I'm rather sceptical that any systematic philosophical account of mathematics in applications will be forthcoming. The best we may ever be able to do is understand particular applications on a somewhat ad hoc and case-by-case basis. But this, of course, is mere speculation.

5 *Empiricism revisited*

So with Leng's argument that all mathematics is recreation dispensed with, we are able to maintain the recreational–non-recreational distinction. This distinction is important for the kind of empiricism I'm advocating here. Even though it might turn out that there is not a great deal of recreational mathematics (if any), there must be room for such activity. For otherwise the empiricism is rather vacuous. We do not want mathematics to be justified simply because some mathematicians study the area in question.[25] That would not be empiricism at all.

I have a great deal of sympathy with the idea that mathematics should be justified on purely mathematical grounds. After all, mathematics is the queen of the sciences and as such might be thought to occupy a privileged position, not in need of any further (external) justification. This view, however, leads to problems. Taking the lead from mathematics, practitioners in other areas might seek justification for their beliefs in terms of their own practices. We might find a push to justify religious beliefs because they belong to a system studied by some religious group or other. Or perhaps an attempt to justify beliefs about extraterrestrial abductions because some UFO cult takes such abductions seriously and claims to study them. Clearly mathematics enjoys a higher status and is much more reputable than either religion or alien abduction theory, but what is it that gives mathematics such status? Empiricism gives a clear answer to this question (at least for all mathematics that's applied): it is justified by its direct and indirect applications in empirical science. Indeed, according to this version of empiricism *all* beliefs must ultimately be answerable to empirical evidence. We are thus able to provide a satisfying account of mathematical knowledge, where mathematics is respected, but it earns this respect by the work it does in empirical science. There is no room for free riders or self-indulgent queens. Everyone pays their way in this version of empiricism—even royalty like mathematics.[26]

[24]See Bueno and Colyvan (n.d.) for some tentative steps towards such an account.

[25]Maddy (1992) suggests extending Quinean naturalism to pay due respect to mathematical practice along such lines.

[26]I'd like to thank Alan Baker, Daniel Isaacson, Mary Leng, Aidan Lyon, Gideon Rosen, and Crispin Wright for very helpful conversations on the topic of this chapter, and Mary Leng for her comments on an earlier draft. I'd also like to thank others in the audience at the Mathematical Knowledge Conference held at the University of Cambridge in June–July 2004 for their comments and criticisms. Finally, I'd like to thank the organizers, Mary Leng, Alexander Paseau, and Michael Potter, for putting together such a terrific conference and for inviting me to speak at it. Work on this chapter was funded by an Australian Research Council Discovery Grant (grant number DP0209896).

Scientific Platonism

ALEXANDER PASEAU

1 *Introduction*

Ask a typical mathematician whether the truth of mathematical statements and the metaphysics of mathematics are settled by natural science, and the answer is likely to be that they are not. He or she might add that many questions of mathematical interest are connected to or originally arose from applications in science, but that the applications of mathematics to science are derivative and that they do not settle its truth or metaphysics. Ask a typical post-Quinean analytic philosopher the same question, and the answer might well be that the truth and metaphysics of mathematics can *only* be settled by considering its scientific applications. I stereotype of course: there are plenty of mathematicians and analytic philosophers who think otherwise, and plenty more who have no opinion to speak of. But such are the broad tendencies. Who is right?

The two questions must be distinguished. First, does natural science give us reason to believe that mathematical statements are true? Second, does natural science give us reason to believe in some particular metaphysics of mathematics? My argument here will be that a negative answer to the second question is compatible with an affirmative answer to the first. Loosely put, even if science settles the truth of mathematics, it does not settle its metaphysics. One epistemological implication is that a scientific defence of our knowledge of mathematical truths need not amount to a defence of our knowledge of mathematical objects.

2 *Preliminaries*

Talk of science settling the truth or metaphysics of mathematics is loose and needs to be cashed out more precisely. We begin by stating the thesis to be defended and then spend most of this chapter's first half clarifying, motivating, and situating it. The second half defends the thesis against some objections.

Let scientific platonism provisionally be the doctrine that scientific standards endorse platonistically interpreted mathematics. The objection to scientific platonism to be articulated is that there is a gap between scientific endorsement of truth-value realism about mathematics and scientific endorsement of platonism. Even if scientific standards endorse the truth of mathematics under some interpretation, I argue, they might not endorse its platonist interpretation. If I am right, a stronger claim is in fact true: scientific standards do not endorse *any* particular interpretation of mathematics. Thus I shall defend the second of the typical mathematician's views, that the nature of mathematics is in this sense independent of natural science, assuming for the sake of argument that science endorses the truth of mathematics.

Platonic or abstract entities are here understood in broad brush terms, as entities in neither space nor time. By 'science' we mean throughout natural science (physics, chemistry, biology, etc.). The expressions 'scientific standards', 'scientific grounds', 'scientific reasons', 'scientific norms', etc., are taken as synonymous, as are 'endorsement', 'recommendation', etc. To say that scientific standards endorse (or recommend) p or that astrological standards endorse q is not thereby to endorse p or q oneself, but merely to point out that these propositions are supported by these standards. Thus to claim that q is endorsed on astrological grounds—because, say, astrologers infer q from the planets' alignment—is not to claim or imply that anyone should accept that as a reason for believing q. What I call scientific standards' endorsement of proposition p could in a different idiom be expressed by saying that p is scientifically confirmed. Endorsement by scientific standards is thus simply scientific confirmation.

Scientific standards are the standards underlying theory evaluation in the natural sciences. Empirical adequacy—agreement with empirical data[1]—for example, is a scientific standard, indeed the paradigm scientific standard, whereas compatibility with the sayings of some sacred person or text is not. Some form of the principle of simplicity, which in its most general version states that the simpler of two theories enjoys a theoretical advantage over the less simple one (in this respect), is also a scientific principle. This follows from generally appreciated facts about scientific practice, for example from the fact that a complicated epicycle theory is generally thought scientifically inferior to one positing a more uniform trajectory, even if the former has been tweaked so as to agree with all existing data. To take another example, scientific standards also recommend unifying theories that account for different groups of phenomena and theories in terms of the same mechanisms, a classic instance being the unification of terrestrial and celestial mechanics in Newtonian mechanics.

Not all standards are scientific. Religious standards are often non-scientific: many religious claims, e.g. that there is an afterlife, or that God directly intervened to cause a tsunami, etc., are endorsed by the standards of various religions but not by scientific standards.[2] Philosophy also serves up some self-conscious appeals to non-scientific standards. Goodman and Quine for example famously

[1]In practice, of course, most actual theories fail to mesh with all the empirical data—they contain 'anomalies'—even from their inception. 'All theories, in this sense, are born refuted and die refuted,' famously wrote Lakatos, because they contain 'unsolved problems' and 'undigested anomalies' (1978: 5). The better confirmed the theory, the fewer these anomalies and the easier they are to explain away.

[2]Some of the claims could potentially be justified by scientific standards (e.g. these standards might recommend, following scrupulous verification of her other predictions, the infallibility of a prophet who claims there is an afterlife); but my point is that, as a matter of fact, such claims typically aren't so justified.

begin their 1947 nominalist manifesto by declaring that the basis for their nominalism is a fundamental 'philosophical intuition' irreducible to scientific grounds.[3]

It is not always clear whether something is a scientific standard. Many philosophers casually (and some not so casually) say that ontological economy is a scientific principle. But the scientific status of the (absolutely) general principle of ontological economy, understood as the claim that any theory with an ontology smaller than that of another theory enjoys a theoretical advantage over it (in this respect), is controversial.[4] In particular, the highly general form that ontological economy takes in the hands of philosophers who invoke it (along with other reasons) to defend, say, resemblance nominalism over a trope theory of properties, or a trope theory over a universals theory, or some particular 'ersatz' modal realism over Lewisian realism in the philosophy of modality, or nominalism over platonism in the philosophy of mathematics, or some regularity theory of natural laws over full-blooded nomological realism, or a semantics based on sentences rather than propositions, etc., *looks* very different from the more local and apparently empirically grounded form it takes in the hands of scientists. Observe in particular that even the slightest difference between the weighting ontological economy is given in cases of philosophical theory choice and its weighting in typical scientific contexts constitutes a divergence from scientific standards. Indeed, it is the customary complaint of platonists such as John Burgess and Gideon Rosen that many philosophers give greater weight to ontological economy than is scientifically proper.[5]

The claim that there are scientific standards should not be confused with the claim that there is a scientific method, in the sense of a procedure which can or should in practice be used to develop new or better scientific theories. A scientific method is a kind of recipe, whereas scientific standards are evaluative.

Whether or not scientific standards are the same as the evaluative standards of other areas of inquiry, such as the social sciences, the humanities, mathematics, etc., is a deep and important question, but tangential here. A more relevant question concerns intrascientific differences. The scientific platonist assumes what Russell meant when he said that 'men of science, broadly speaking, all accept the

[3] 'We do not believe in abstract entities ... We renounce them altogether ... Fundamentally this refusal is based on a philosophical intuition that cannot be justified by appeal to anything more ultimate' (1947: 105). Goodman and Quine go on to add that this fundamental rejection is fortified by certain a posteriori considerations. Note that Quine in his later, more famous incarnation (which serves as the inspiration for contemporary scientific platonism) repudiated this intuition.

[4] Different understandings of what it is to have a small ontology result in different versions of this principle. (For example, one might distinguish between quantitative and qualitative economy.) Burgess (1998) argues that scientific standards do not endorse theories with smaller *abstract* ontologies. If he is right, it follows that the fully general principle of ontological economy is not scientific. Note that many philosophers call a principle along these lines Ockham's Razor. I find this usage unhelpful, because Ockham's Razor is the standard label for the principle that entities should not be multiplied beyond necessity. Without further elaboration, that is a platitude.

[5] 'The reconstructive nominalist [the philosopher who seeks to reconstruct science on nominalist lines] seems to be giving far more weight to ... economy, or more precisely, to a specific variety thereof, economy of abstract ontology, than do working scientists. And the reconstructive nominalist seems to be giving far less weight to ... familiarity and perspicuity' (1997: 210).

same intellectual standards' (1945: 599). It is not entirely obvious, however, that the natural sciences all employ the same standards. It is not entirely obvious, for example, that biologists and physicists apply the same set of standards, that is, that they use the same set of criteria by which to evaluate their theories. Still less obvious is it that scientific standards have remained fixed throughout history, even following the scientific revolution of the seventeenth century.[6] It is unclear how troubling, if at all, it would be for the scientific platonist to peg her thesis to the scientific standards of some specific era—perhaps she should simply peg it to the standards of 'our era'. More troubling is the prospect that, even at a given time, there might be no such thing as global scientific standards, but just the standards of this or that part of science, or even just the standards of this or that group of scientists. Substantial though they be, let us go along with the scientific platonist's assumptions here in order to give the view a run for its money. In keeping with the assumptions, I shall use the monolithic label 'scientists' for the best (though not infallible) deployers of scientific standards, slurring over distinctions between them.[7]

Scientific standards can of course endorse propositions to a certain degree or with certain qualifications rather than outright. I ignore such complications as they do not affect the discussion.

To say that scientific grounds endorse some theory is roughly to say that scientists correctly endorse it *qua* scientists. Thus scientific grounds could endorse p even if all scientists disbelieve p for scientific reasons (they might all be mistaken in their scientific evaluations), or even if each scientist disbelieves p all things considered (e.g. 'scientific grounds unequivocally support p but my overriding religious convictions tell me not-p'). It would be nice if there were a reductive analysis of correct endorsement of a proposition *qua* scientist; but to my knowledge none exists. The thesis of scientific platonism is hardly the worse for it.

Endorsement is understood epistemically. If scientific grounds endorse platonism merely as a useful but ultimately false assumption, like, say, the useful but false assumption of the uniform density of some fluid, the conclusion, if we accept scientific standards, would be not that we should believe platonism, but (at best) that it is a useful but false hypothesis. Scientific platonism will accordingly turn out to be false if scientific grounds should recommend acceptance of platonist mathematics only in some non-epistemic sense. Following standard usage, I call any such non-epistemic sense of support for a proposition p 'pragmatic' support for p.

We should be clear about the difference between endorsement of a particular interpretation of some sentence and endorsement of the proposition expressed

[6] Larry Laudan is a philosopher known for arguing the opposite: 'The specific and "local" principles of scientific rationality which scientists utilize in evaluating theories are not permanently fixed, but have altered significantly through the course of science' (1981: 144).

[7] The intended sense of 'scientist' is thus not a narrow institutional one: to qualify as a scientist one need not be employed in a university or a scientific research institution (even if these days most scientists are).

by that sentence under some interpretation. Suppose that history is right in relating that Heraclitus once said 'πάντα χωρεῖ καὶ οὐδὲν μένει'.[8] Proper standards of translation endorse translating Heraclitus as saying (roughly) 'all is in flux and nothing stays'; but it does not follow that these standards endorse the claim that nothing stays and that all is in flux. Likewise, when we say that scientific standards endorse the platonist interpretation of some sentence s, we mean that scientific standards endorse the proposition expressed by s interpreted platonistically, not that scientific standards endorse the claim that the platonist interpretation is the right one to put on the claim's standard utterances or inscriptions. Endorsing p is quite different from endorsing the claim that an utterance of sentence s expresses proposition p.

The question of what follows if scientific standards endorse platonism is a fundamental epistemological one. Some scientific platonists have gone as far as to say that if scientific standards endorse p then we should believe p. This is a fairly extreme form of epistemic naturalism or scientism, which takes scientific standards to trump all others (the most extreme kind of naturalism would strengthen the conditional into a biconditional). Less extreme views give scientific standards some weight without making them the ultimate authority. The lesser the weight, of course, the less significant the scientific platonist's thesis. What motivates this paper is initial sympathy with the view that scientific standards have some say in the philosophy of mathematics. Exploration of logical space for its own sake has its place; but I for one would lose my interest in the debate if it became clear that scientific standards should be accorded no weight whatsoever, as the stereotypical pure mathematician would have it. Be that as it may, our discussion can float free of the difficult epistemological issue of how much weight to accord scientific standards in the evaluation of a philosophy of mathematics.[9]

Scientific platonism is obviously inspired by the Quine–Putnam indispensability argument. The argument, roughly, is that platonism is true because platonist mathematics is indispensable to our best scientific theories. Implicit within this line of thought is the naturalist premise that we had better believe the deliverances of our best scientific theories, whatever the non-scientific reasons to the contrary. The second premise of the indispensability argument is that our best science indispensably contains platonist mathematics. The literature of the past few decades, however, has suggested that a non-platonist mathematics may be developed for the purposes of science (see below). If this is right, then platonist mathematics is not in principle indispensable to science. The reply by scientific platonists has been that even if a non-platonist mathematics could be developed, it would still be inferior to platonist mathematics by scientific standards (e.g. this is the main thesis of Burgess and Rosen (1997)). The question thus becomes one of scientific superiority rather than indispensability in the strict sense, and the indispensability question accordingly turns into the question of whether scientific

[8]Plato, *Cratylus*, 402a8–9.

[9]For discussion of some of the issues here, see my (2005).

platonism is true. (Arguably, 'indispensability' was understood in this way from the start, despite the choice of word.)[10]

Quine's influence being what it is, particularly in the United States, scientific platonism has plenty of adherents. Contemporary Quineans include Alan Baker, John Burgess, Mark Colyvan, Michael Resnik, and Gideon Rosen among others.[11] It might be thought somewhat misleading to call these and other contemporary Quineans scientific platonists. After all, several of them follow Quine in claiming that philosophy is continuous with science, and they might in principle baulk at the claim that scientific standards endorse platonism if this is taken to imply a sharp demarcation between natural-scientific standards and philosophical ones. In practice, however, they are happy to speak of scientific standards as relatively well-demarcated. For instance, John Burgess and Gideon Rosen in a section of their 1997 book list the standards generally accepted by descriptive methodologists of science as scientific and use this as a club with which to beat nominalists, accusing them of producing reconstructions that are scientifically inferior when judged by these very standards ((1997: 209 ff.); see also (2005: 519–20)). Baker (2001: *passim*) is another example of someone who speaks of 'scientific grounds' without hesitation. In general, theoretical caution about the perceived or potential continuity of scientific grounds and philosophical grounds does not prevent Quineans from espousing scientific platonism. The scientific platonist will typically concede that there is no sharp distinction between scientific and non-scientific grounds for belief, and no sharp demarcation of science, but she will nevertheless insist that this does not undermine her position.

A potentially stronger reason for not attributing a blanket scientific platonism to the mentioned writers is that some of them understand 'science' and 'scientific' in a broad sense that goes beyond the natural sciences. For instance, if one counts mathematics as part of science, one could say that 'scientific' standards in this sense endorse platonism simply because mathematical standards endorse platonism and natural-scientific standards do not speak against it. This seems to be John Burgess's view at the time of writing, and it might be the official

[10]For a succinct version of the indispensability argument by Putnam, see his (1971). For Quine, see his famous (1953) and many of his later writings (e.g. the articles in his (1981*b*)). Resnik (2005) offers an accessible overview of Quine's philosophy of mathematics. Note that Quine often construes science in a broader sense than the one here (see, e.g., his (1995: 49)). Some philosophers have recently claimed that a version of the indispensability argument can be found in §91 of Frege's *Grundgesetze*. Garavaso (2005) argues that this attribution is mistaken.

[11]Colyvan ((2001) and this volume), Resnik (1997); Burgess and Rosen (1997); Baker (2001). For exactly how to interpret Burgess and Rosen (1997), see nn. 12 and 26 below. Resnik's version of the indispensability argument consists of the following two premises (by 'science' he understands natural science): '(1) In stating its laws and conducting its derivations science assumes the existence of many mathematical objects and the truth of much mathematics. (2) These assumptions are indispensable to the pursuit of science; moreover many of the important conclusions drawn from and within science could not be drawn without taking mathematical claims to be true.' (1997: 46–7) Colyvan's platonism is discussed below. Baker's platonism stems from his belief that scientific grounds support platonism, together with the credo 'that—given the naturalistic basis of the Indispensability Argument, which rejects the idea of philosophy as a higher court of appeal for scientific judgments,—the only sensible way of judging alternatives to current science is on scientific grounds' (2001: 87).

view of Burgess and Rosen (1997).[12] This kind of 'scientific' platonism might be more accurately termed *mathematical-cum-scientific* platonism. Mathematical-cum-scientific platonism is less controversial than scientific platonism proper, since it is often thought that mathematical standards endorse platonism and that the philosophical project begins when we start asking whether there is a better all-things-considered account of mathematics. From the perspective of the mathematical-cum-scientific platonist, who thinks that mathematical standards endorse platonism, a discussion that focuses only on (natural-)scientific standards will accordingly be seen as conceding too much ground to the anti-platonist. Conversely, from the perspective of a philosopher who takes only natural scientific standards as justificatory, anyone who believes in platonism because it is sanctioned by mathematical standards but not necessarily scientific ones will be seen as begging the question on behalf of platonism. However that may be, we should be clear that mathematical-cum-scientific platonism is quite distinct from our present quarry, *viz.* (natural-)scientific platonism. The least we can say is

[12]It is unclear whether Burgess and Rosen (1997) wish to defend scientific platonism or mathematical-cum-scientific platonism or both. Pages 32–5, for instance, seem to point to mathematical-cum-scientific platonism, perhaps augmented by common sense. (E.g. 'Another form of objection questions whether there is any viable notion of "justification" other than that constituted by ordinary common sense and scientific and mathematical standards of justification' (p. 32; see also p. 211).) Elsewhere, however, Burgess and Rosen simply speak of scientific standards. (E.g. 'The naturalists' commitment is at most to the comparatively modest proposition that when science speaks with a firm and unified voice, the philosopher is either obliged to accept its conclusions or to offer what are recognizably scientific reasons for restricting them' (p. 65; see also p. 205).) And the fact that Burgess and Rosen do not define science somewhat non-standardly as comprising the natural sciences *as well as* mathematics, and moreover that they use the words 'scientific' and 'science' *in contrast* to 'mathematical' and 'mathematics'—as illustrated by the quotation from p. 32 and in many other passages—supports interpreting them as scientific platonists and not mathematical-cum-scientific platonists. (Of course there is a tradition of calling mathematics and logic the 'formal sciences', but the default contemporary understanding of 'science' is to denote the natural sciences.) Likewise, saying that correctness and accuracy of observable predictions are among the standards that descriptive methodologists agree are operative in science (p. 209) is suggestive of the natural sciences, as these two standards are not relevant to mathematics, at least not in any literal sense. This unclarity runs throughout their book and is unfortunately never resolved. (It resurfaces for instance in their discussion of the publication-and-reception test, for more on which see n. 26 below.) Perhaps the most reasonable interpretation is to take the book as committed to both theses. Another terminological minefield is John Burgess's self-labelling. Burgess has in recent years taken to calling himself an 'anti-anti-realist' rather than a platonist (see e.g. his (2004)). An anti-anti-realist is said to be someone who does not take back in the philosophy seminar what he says in the mathematics classroom. But since Burgess also thinks that what is said in the mathematics classroom is true, and is intended literally, and that this literal truth entails platonism, it seems to follow that his anti-anti-realism is just a version of platonism. In the time-honoured tradition of burdening one's opponent with an inflated version of his actual position, Burgess seems to think, however, that to be a platonist is to be a 'capital-R-realist'. The principal non-metaphorical definition of this character Burgess offers is that a capital-R-realist thinks that 'what one says to oneself in scientific moments when one tries to understand the universe corresponds to Ultimate Metaphysical Reality' (2004: 19), a claim which Burgess disowns. Burgess's distinction between capital-R-realism and anti-anti-realism (for more on which, see pp. 34–35 of his 2004) is difficult to understand, however. For one thing, it is difficult to see how metaphysical reality itself could come in various forms, ultimate, penultimate or preliminary: what is the case is simply the case, *c'est tout*. For another, Burgess's distinction between thinking that a statement is true and thinking that it corresponds to reality (in a sense that allows of course that other intelligent beings could correctly conceptualize the world differently from us) is unclear. All in all, even if he dislikes the label, there are good reasons for eliminating his coy double negations and for calling Burgess a platonist.

that scientific platonism is typically accepted by most contemporary Quineans—indeed, it is almost a definition of being Quinean in this domain—and that for them it usually forms the central plank of their case for platonism, just as it did for Quine. In short, even if you think there are justifications for platonism *other* than scientific ones, if you think that scientific standards justify platonism, that makes you a scientific platonist.

The unavoidably lengthy preliminaries over, I now develop an objection to scientific platonism in sections 3–5 based on the idea that scientific grounds are indifferent between platonism and interpretations equivalent to it (in a sense to be explained). Sections 6–8 buttress that objection by replying to some counter-arguments.

3 *The pragmatic and indifference objections*

A two-step line of thought leads to scientific platonism. The first step is that to accept a mathematical truth such as '2 is prime' seems to commit one to the existence of a mathematical entity (the number 2) and to a mathematical property (being prime). This literal reading of mathematics—call it *realism*—is the scientifically assumed one, and is therefore scientifically warranted. (NB realism as here understood makes no claims about the nature of these entities.) The second step takes us from realism to platonism, the claim that the objects of which mathematics speaks when correctly construed at literal face value are abstract. This second step is supposed to follow from the fact that scientists (along with everyone else) implicitly understand that mathematical entities are not concrete. Scientific standards may not condone any definite positive conception of the entities posited by scientifically applied mathematics. But they support at least a negative characterization: these entities, whatever they are, are not concrete.

The assumption of scientific endorsement contained in this line of reasoning has recently been questioned. Penelope Maddy has argued that close attention to scientific practice suggests that 'the success of a theory involving certain mathematical existence assumptions and corresponding physical structural assumptions is not regarded as confirming evidence for those assumptions' (1997: 156). For instance, she maintains that even though scientists standardly employ the hypothesis that space–time is continuous, they do not think there is compelling evidence for it. She argues more generally that 'in some cases, a central hypothesis of an empirically successful theory will continue to be viewed as a 'useful' fiction until it has passed a further, more focused, and more demanding test' (1997: 142). Mathematical existence claims, according to her, often fail or are not subjected to this latter kind of test. Hence her conclusion, that scientific grounds strictly speaking do not endorse platonist (or even realist) mathematics.[13]

The claim that scientific grounds do not endorse the truth of platonistically interpreted mathematical statements may be broken down into two disjunctive

[13]For more detail on Maddy's views, see her (1992), (1995) and (1997: ch. II.6).

components. One is that scientific grounds do not endorse the truth of mathematical statements. The other is that the statements whose truth they endorse are not platonist. The rest of this section expands on the difference between these two claims and their associated objections. (We return to the first paragraph's reasoning in section 8.)

According to the first objection, which I call the *pragmatic objection*, scientists (*qua* scientists) are not epistemically committed to the mathematics they deploy. Science does not endorse mathematics in an epistemic sense, but at best in a pragmatic one. Modern science stripped of mathematics would of course be highly impoverished, indeed more or less unrecognizable, but according to the pragmatic objection that does not mean that the mathematics that is part of science is thereby confirmed. In particular, confirmatory holism is not true, since the mathematical portion of a well-confirmed scientific theory is not necessarily confirmed. Indeed, according to the pragmatic objection it is *not* confirmed. If sound, this objection would be sufficient to defeat the scientific platonist, since the latter aims to establish belief in, rather than pragmatic acceptance of, platonist mathematics.

By contrast, the *indifference objection* maintains that scientific standards endorse mathematics in the proper epistemic sense but that they do not endorse *platonist* mathematics. Let S^P be science together with platonically interpreted mathematics and let S^{NP} be science together with some non-platonist interpretation of mathematics that makes the same claims about the physical world as S^P, assuming for now that some such S^{NP} exists. The indifference objection to the argument's second premise is that scientific grounds do not endorse S^P (or S^R) over any such S^{NP}.[14]

The difference between the two objections is important but subtle, so is worth highlighting. The pragmatic objection argues that scientific standards do not endorse the mathematics applied in science in the proper epistemic sense. (The endorsement is, at best, merely pragmatic.) The indifference objection concedes that scientific standards endorse the mathematics applied in science in the proper epistemic sense, but it urges that the mathematics thereby endorsed is not platonist (nor realist). More strongly, it contends that there are no scientific grounds for distinguishing between the apparent scientific equivalents S^P and S^{NP}. According to the indifference objection, science's epistemic endorsement is of the truth of mathematics under some acceptable interpretation, but not any specific one.

Unscrambling the two objections is important for the sake of clarity. But it is particularly important in light of the fact that the recent debate has focused on the

[14] S^R is (natural) science together with realistically interpreted mathematics. The notation suggests that there is only one platonist interpretation, but of course there are many, as briefly mentioned at the end of this section. Note that I am implicitly restricting attention to non-platonist interpretations that make the same claims about the physical world as the platonist interpretation: a non-platonist interpretation that, say, identifies numbers with physical objects is therefore excluded. Note further that scientific standards rule out certain unacceptable interpretations such as 'mathematics is all false but if the platonist were right then it would be the case that...'; S^{NP} is therefore restricted to acceptable interpretations throughout.

pragmatic objection and ignored the indifference objection. Maddy's arguments are most naturally interpreted as versions of the pragmatic objection. The claim, for instance, that the use of mathematics in science is analogous to the use of not-believed-to-be-true (or even known-to-be-false) idealizations and assumptions in science implies that scientific standards do not endorse the truth of mathematics. Another noted critic of scientific platonism, Elliott Sober (1993*a*), has also argued that mathematics does not receive empirical confirmation, his main objection being that mathematics is apparently never disconfirmed by empirical *failure* (in a case of failure it is the science, not the mathematics, that gets blamed) and hence that mathematics should not be seen as confirmed by empirical *success* either. As he is also sceptical about non-empirical sources of confirmation (1994*a*; 1994*c*), Sober is therefore a proponent of the pragmatic objection. Despite its popularity in some quarters, however, this objection remains controversial—see Mark Colyvan's contribution to this volume for some responses to it. Developing the indifference objection therefore provides anti-platonists with a second, perhaps stronger, line of attack against scientific platonism, which may succeed even if the first fails. Anti-platonists may therefore base their case disjunctively on both objections and avoid putting all their eggs in one basket.

The indifference objection is to be distinguished from the claim that science is indifferent between standard mathematics and the segment thereof that finds scientific application. Even a relatively weak mathematical theory (much weaker than, say, standard set theory, ZFC) will probably do, at least in principle, for all scientific purposes.[15] The claim that scientific grounds only endorse some of current mathematics (e.g. anything below 'higher' set theory, on some way of demarcating it), however, is very different from that mooted here. The indifference objection is not committed to making a controversial distinction *within* mathematics between its scientifically and non-scientifically confirmed parts, the latter being regarded as mathematical recreation. It is compatible with the objection that science endorses mathematics wholesale, even if it does not endorse any particular interpretation. The indifference objection does not in itself draw a line across mathematics separating the part that is scientifically confirmed from the rest.

The indifference objection is also different from the claim that scientific standards endorse platonism but not any specific version thereof (e.g. set-theoretic, category-theoretic, property-theoretic, etc., platonism). This is thought by some (Wagner 1996; Baker 2003) to be a limitation of the scientific argument for platonism. Steven Wagner expresses the charge succinctly:

> A limitation of the argument from science is that it leaves the abstract ontology indeterminate. Any abstract ontology that works will admit countless alternatives. Our numbers can be properties, properties sets, and sets numbers; our pure sets can be impure; ordered objects can be construed as unordered ones or vice versa; and so on. Science seems to have

[15] See, for instance, Feferman (1998: part V).

the curious feature of requiring a substantial abstract ontology but none in particular. (1996: 80)

The complaint that scientific standards do not endorse a specific platonist ontology for mathematics clearly arises at a later stage than the indifference objection. One has first to accept that scientific standards endorse platonist mathematics even to consider whether they endorse a specific kind of platonism. The indifference objection urges that we cannot even go that far.

4 *Weak and strong scientific platonism*

What I have called the indifference objection is really composed of two theses corresponding to its first and second words. The 'indifference' part is that scientific standards do not endorse S^{P} over (any acceptable) S^{NP} and vice-versa. The 'objection' part is that some such S^{NP} exists that is not committed to abstract objects.

Non-platonist interpretations of mathematics abound, two recent examples being Geoffrey Hellman's modal-structuralism and David Lewis's structuralism. Modal-structuralism is an interpretation of mathematics mathematically equivalent to platonism.[16] Roughly, a claim p is interpreted as the claim that its structuralist analogue p^{S} necessarily holds in any structure that instantiates the axioms of the branch of mathematics in which p features.[17] Thus the interpretation of p is the necessitation of the universal Ramsification of p conditional on the axioms of the relevant branch of mathematics. A simpler version of structuralism, offered by David Lewis (1993; 1991), does away with the modal operators and posits the existence of enough entities to provide a model for standard set theory (and thus for all branches of mathematics). S^{NP} is more generally science together with some such non-platonist interpretation of mathematics.

To give readers a little more to sink their teeth into, let me expand briefly on Hellman's modal-structuralism as applied to arithmetic. Very roughly, Hellman interprets an arithmetical claim p as the claim that its structuralist analogue p^{S} necessarily holds in any structure that instantiates the axioms of Peano arithmetic. A semi-formal statement of the modal-structuralist interpretation of, say, '$0 \neq 1$' would be, 'Necessarily, for any collection of elements X, for any function S_X on X and element 0_X of X, if the Peano axioms hold of (X, S_X, 0_X) then $0_X \neq S_X(0_X)$'. Hellman (1989: 47–52) then argues that the logical apparatus required for his theory (e.g. second-order quantifiers) does not commit him to abstract objects. This is of great importance to him, since he would like 'to leave open the possibility of a "nominalist" reading of the mathematical theories

[16]See Hellman (1989) for the book-length treatment. A more recent exposition may be found in Hellman (2005: 551–60).

[17]A background assumption must also be added that such a structure is possible. The case of set theory (in particular, unbounded set-theoretic sentences) requires special handling (Hellman 1989: 73–9).

in question' (1989: 20 n. 11; see also 47–52, 105–17). It is equally important that a translation scheme (whose outline is straightforward but whose details need not detain us) should exist between realistically construed arithmetic and his modal-structuralist interpretation. As he explains,

> Recovery of proofs is, however, only the first step in justifying the translation scheme. As already emphasized, the modal-structuralist aims at much more: in some suitable sense, the translates must be mathematically equivalent to their originals. (1989: 26)

Most of the lengthy first chapter of (1989) is taken up with explaining the sense in which this proffered translation scheme provides a mathematical equivalence. The equivalence is of course not a logical one; otherwise Hellman would have to concede that a realist statement p^{R} is true if and only if its modal-structural counterpart p^{MS} is true, yet the whole point of his approach is that non-realist statements might be true even if realist ones are not. So the rough idea is that p^{R} and p^{MS}, although not logical equivalents, and not even truth-conditional equivalents (since one could be true and the other false), are associated by paraphrase and share the same inferential properties within their associated networks.[18]

I should emphasize that our sense of 'interpretation' is different from the standard model-theoretic one in which an interpretation is specified by fixing a domain of entities and a compositional function from a formal (uninterpreted) language to elements of the domain and set-theoretic constructions thereof. Hellman's modal structuralism, Lewis's non-modal structuralism, realism and platonism all count as interpretations of mathematics in our sense but not in the model-theoretic one. What we might call a Reinterpretation Function takes sentences of the body of accepted mathematics interpreted in one way (e.g. realism) to sentences interpreted in another (e.g. modal-structuralism). This function satisfies a compositionality constraint, is recursive on the language (of each branch of mathematics) and of course respects intended truth-values. Though it is worth detailing the Reinterpretation Function's further properties (in particular, its exact domain and range), this programmatic paper is not the place for it.[19]

As I am not sympathetic to formalism, I take mathematics as actually written or spoken to be a collection of meaningful sentences (or propositions) rather than uninterpreted ones. My use of locutions such as 'platonistically interpreted mathematics' is therefore not meant to suggest that mathematics has to be interpreted to be meaningful; rather, the phrase designates the collection of mathematical sentences understood as the platonist understands them. The term 'interpretation', I believe, nicely focuses what is at stake between, say, platonist and structuralist accounts of standard mathematics; but it should not mislead us into thinking that standard mathematics is uninterpreted. In particular, the natural (though perhaps

[18]Some writers, such as Putnam (1983), see realist and structuralist interpretations as truth-conditionally equivalent, but this is the exception rather than the rule.

[19]As pointed out in n. 14, not every interpretation that meets these minimal conditions is acceptable by scientific standards.

naive) view that standardly understood mathematics is realist is compatible with everything in this paper (see section 8).

Earlier, we provisionally defined scientific platonism as the claim that scientific standards endorse the platonist construal of mathematics. Taking our cue from the two parts of the indifference objection, we now settle upon our final terminology and relabel this thesis *strong scientific platonism*. According to strong scientific platonism, then, scientific grounds endorse the platonist interpretation of mathematical statements such as, say, '$\exists x \exists y(x \neq y \wedge x \in y)$'—they endorse the claim that there are two abstract objects, the first of which has the abstract property of being a member of the second. According to *weak scientific platonism*, by contrast, scientific standards endorse the claim that mathematics is committed to the existence of some platonic entities. Obviously a commitment to a platonist reading of '$\exists x \exists y(x \neq y \wedge x \in y)$' would amount to a commitment to a platonic entity. But another way such a commitment might be incurred would be if, for example, scientific standards endorsed the structuralist interpretation of set theory *and* the claim that there are insufficiently many concrete entities for the structuralist reading to deliver the right truth-values of set-theoretic statements. Strong scientific platonism therefore implies weak scientific platonism but not the other way round. As a matter of fact most past scientific platonists have been strong scientific platonists.

Now the indifference objection, if true, immediately defeats strong scientific platonism. Whether or not it defeats weak scientific platonism depends on whether (it can be scientifically shown that) there is at least one acceptable S^{NP} not committed to abstract objects. The indifference objection to weak scientific platonism is therefore based on the important assumption that there is at least one such S^{NP}.

To appreciate the assumption, consider whether any form of structuralism must ultimately be committed to abstract objects (perhaps not distinctively mathematical ones). The case of modal structuralism's ultimate ontological commitments is currently moot, revolving on the acceptability of its primitive modal ideology.[20] As for non-modal structuralism, a continuous space–time and its regions offer a model of second-order real analysis.[21] Second-order analysis, however, is apparently sufficient for all current scientific applications of mathematics; and space–time regions are arguably concrete.[22] Hellman (1999) argues that (full classical) fourth-order number theory (or third-order real analysis) can be captured within his modal-structuralist framework (supplemented with mereology and plural quantification) without any overall commitment to abstract entities.

[20]For some points on both sides of the debate, see Hellman (1989) and Shapiro (1993).

[21]Space–time points constituting the domain of first-order quantifiers, and regions (construed as mereological sums) constituting the domain of the second-order quantifiers. Some fiddling is required to account for the empty set.

[22]They arguably have spatiotemporal locations, although it might be unnatural to say they are *in* space–time.

When it comes to set theory and other branches of mathematics that go beyond this framework (e.g. category theory, some parts of functional analysis, algebraic topology, etc.), an ontology of Lewisian concrete possibilia would suffice for a model of its structuralist version. Determining whether an appropriate infinity of concrete possibilia exists evidently involves plunging into deep metaphysical waters. But to acknowledge as much is to acknowledge the current status of these claims. Like most people, I am not particularly sympathetic to Lewisian realism about possibilia. Nevertheless, I recognize that it is hardly straightforward to show that a structuralist construal of mathematics is committed to abstract objects, if this involves refuting Lewisian realism in the bargain. Indicting structuralism of commitment to abstract objects is not that easy. And indicting structuralist second-order analysis of ultimate commitment to abstract objects is even harder. Moreover, it is dubious whether these charges could be upheld on the basis of a scientific argument. Finally, the claim that no metaphysics of concreta can supply a structuralist set theory's ontology is overweeningly general. Establishing it would require some capacity to enumerate or capture the relevant general features of any such metaphysics; but of course it is doubtful that we can currently do so. So it is fair to say that the jury is still out. Since I do not want to tackle the question further here, I shall continue to proceed conditionally, having indicated that it is not obvious that such an S^{NP} does not exist.

To summarize: the indifference objection, if sound, sinks strong scientific platonism. Whether it also sinks weak scientific platonism depends on whether *every* (acceptable) S^{NP} scientifically equivalent to S^{P} is ultimately committed to abstract objects. On the one hand, there is no acceptable S^{NP} universally agreed to be free of commitment to abstract objects (some think there is, others not). On the other hand, no half-decent argument is currently available for the sweeping conclusion that there could be no such S^{NP}. A cautious proponent of the indifference objection should currently withhold judgment on whether it defeats weak as well as strong scientific platonism.

5 *For the indifference objection*

One motivation for the indifference objection is that many scientists (at least *qua* scientists) see mathematically equivalent interpretations such as platonism and modal-structuralism as notational variants. Furthermore, most of those who do not do so typically think that the choice between them is scientifically irrelevant. Underlying this is the view commonly held by scientists (and by virtually everyone else for that matter) that mathematics is an auxiliary to scientific endeavour rather than its subject matter. Scientists' judgement that S^{P} (or S^{R}) and S^{NP} are scientifically equivalent (i.e. that scientific grounds do not favour one over the other) constitutes strong evidence on behalf of the indifference thesis. Scientists assume the truth of mathematics yet apparently assume nothing about its metaphysics.[23]

Proponents of the indifference objection may also deploy standard scientific realist arguments, taking in the mathematical parts of science, to support the claim that the best explanation of scientific success invokes the truth of mathematics. We shall not enter this debate here, since we are assuming for present purposes that the pragmatic objection fails. But note that the indifference objection is considerably stronger than well-known objections to scientific realism that proceed by claiming that scientific grounds do not distinguish between observationally equivalent but spatiotemporally inequivalent theories, e.g. between some standard theory T and the sceptical hypothesis that our observations are *as if* T were true though T is in fact false. Since S^{P} and S^{NP} differ only in their interpretation of mathematics, they are spatiotemporally equivalent: they make the same claims about the spatiotemporal world. Now it is one thing to say that the realm of the scientific goes beyond observational adequacy; it is quite another to say that it goes beyond spatiotemporal adequacy.

The indifference objection also captures some of the pragmatic objection's attraction by allowing that scientific grounds recommend S^{P} over many, perhaps all, instances of S^{NP} in a pragmatic sense. The platonist (or realist) interpretation, it might be agreed, is the more convenient one, practically speaking. This is consistent with the thought that scientific grounds do not recommend S^{P} over any S^{NP} in an epistemic sense, which is what the scientific platonist requires. The indifference objection can acknowledge this important pragmatic point while simultaneously respecting the thought that scientific standards (epistemically) endorse mathematics.

Finally, the scientific platonist arguments in the literature appear to be mostly targeted against the pragmatic rather than the indifference objection. A representative example is Mark Colyvan's claim that 'there is good reason to believe that the mathematized version of a theory is "more virtuous" than the unmathematized theory, and so there is good reason to believe mathematics is indispensable to our best physical theories' (2001: 80). Colyvan's argument for this in his book on indispensability arguments, reiterated briefly in his article in this volume, proceeds by 'appealing to a number of examples in which mathematics contributes to the unification and boldness of the physical theory in question, and therefore *is* supported by well-recognized principles of scientific theory choice' (2001: 81). For example, he cites the important role that complex analysis plays in differential equations (relevant to many areas of science), and the importance of the Dirac equation and Lorentz transformation in modern physics. The thing to notice, however, is that Colyvan's counterarguments apply only to the pragmatic objection.[24] If sound, they show that (the truth of) mathematics receives scientific backing; but they fall short of showing that platonist mathematics does. This

[23] Notice that this point in favour of the indifference objection is also a point against the pragmatic objection.

[24] Colyvan (2001) contains extended responses to Maddy and Sober's respective versions of the pragmatic objection.

is true more generally of most defences of scientific platonism—and of course of standard 'indispensability' ones—with a few exceptions to be considered below.[25]

The charge that scientific grounds do not endorse platonist mathematics is therefore a conflation of two very different objections, a fissile compound whose constituents are best treated individually. One is the pragmatic objection, which states that scientific grounds do not *endorse* platonist mathematics, because they do not endorse mathematics. The other is the indifference objection, which states that scientific grounds do not endorse *platonist* mathematics, even if they endorse mathematics. Although the former has received attention, the latter, arguably the stronger of the two, seems to have gone unnoticed. Evaluating it in depth is a tricky task, requiring a detailed study of the role of mathematics in science. To demonstrate that the indifference objection cannot be straightforwardly dismissed, however, I tackle three responses to it in the rest of the paper. Section 6 discusses the idea that there is an easy operational test that will confirm that scientific standards endorse platonism, namely: Let the scientists decide in the usual way! Section 7 considers the idea that general principles of scientific method favour S^{P} (or S^{R}) over any S^{NP}. Section 8 examines the argument that realism is the standard interpretation of what scientists mean by their mathematical utterances and that it is therefore condoned by scientific standards, since these standards condone mathematics' standard interpretation.

6 *The publication test*

Is there an operational test to determine what scientific standards endorse? One might argue for example that whether scientific grounds vindicate nominalism can simply be determined by submitting a nominalist construal of some scientific theory (e.g. Hartry Field's nominalization of Newtonian mechanics) to a scientific journal. If the journal publishes it and it is well received by the scientific community, the answer is 'yes'; otherwise the answer is 'no'. This startlingly simple proposal has in fact been advanced by John Burgess and Gideon Rosen:

> Ultimately the judgment on the scientific merits of a theory must be made by the scientific community: the *true test* would be to send in the nominalistic reconstruction to a mathematics or physics journal, and see whether it is published, and if so how it is received. (1997: 206, my emphasis)

These claims are repeated elsewhere in their book. They maintain that the scientific acceptability of various construals of mathematics and reformulations of science can be gauged by their success in passing this test, and that the test will vindicate platonism.

Despite what they say, however, the test is not a 'true' one. The publication test won't do as some kind of operationalization of scientific platonism (or

[25] Colyvan's assertion that general scientific principles favour platonism over non-platonism (2001: 128–9) does speak to the indifference objection and is considered in section 7.

mathematical-cum-scientific platonism).[26] The reason is that the question of an adequate construal of mathematics is not considered by science journals. The one science journal Burgess and Rosen cite (1997: 210), *Physical Review*, is edited and read by physicists. It is a reasonable bet that few, if any, of these physicists have seriously contemplated any of the issues that preoccupy nominalists. And even if some of the editors have, they would not expect their readers to have done so. And even if, contrary to actuality, many or most—even if you like, all—the editors, referees, and readers of *Physical Review* have pondered these issues, they might still collectively judge it to be an inappropriate forum for such discussion. So the editorial board is likely to reject off-hand any nominalistic submission. But that does not in itself show that the nominalists haven't been addressing a scientific question, albeit a high-level one. Nor does it show that they have erred in the application of scientific standards. What is correct by scientific standards is not co-extensive with what is currently publishable in scientific journals.[27] This is not to say that scientific standards vindicate nominalism, but rather that the presumed unwillingness of *Physical Review* to publish a nominalistic construal of mathematics does not establish its scientific untenability. Failure of the publication test (or a fortiori of the publication-and-reception test) is no touchstone of scientific inferiority.

Another reply would be that the question of nominalism versus platonism is not even a scientific question. If so, rejection of a nominalist submission would be a consequence of the unscientific nature of the question, and not, as Burgess and Rosen assume, of the answer's scientific inferiority. Failure of the publication (or publication-and-reception) test can thus in principle be attributed to two factors other than scientific inferiority. One is that scientists take their trade journals to be inappropriate fora for such discussions, even if they ultimately deem the question scientific. Another is that the question addressed is not scientific, and that scientists recognize this fact. Failing the test is compatible with both alternative explanations.

Evidence for which of the three explanations—the inappropriateness of the forum, the unscientific nature of the question, or the construal's scientific inferiority—is correct could be gleaned from the comments offered by the journ-

[26]Burgess and Rosen's discussion of the publication-and-reception once again illustrates the difficulty of determining whether in their (1997) they uphold a purely scientific form of platonism, or mathematical-cum-scientific platonism, or both. If their platonism were mathematical-cum-scientific rather than just scientific, they would have explicitly cited not just *Physical Review* but also some mathematical journal as appropriate journals to send the reconstruals to. And they would not without expansion or qualification write that 'The [nominalist] innovation, we suspect, simply would not be recognized as progress by practising scientists. And this is so not just for physics, we suspect, but for every natural or social science.' (1997: 210) But the passage quoted above is more suggestive of mathematical-cum-scientific platonism than scientific platonism, since they allow for mathematical as well as scientific journals. So perhaps the most reasonable interpretation is to take them as espousing both. Although I discuss the publication-and-reception test only in relation to scientific platonism, the same points carry over to mathematical-cum-scientific platonism.

[27]We are only considering original, interesting, etc., research—otherwise the publication fails for more banal reasons. And of course we also assume, as Burgess and Rosen intend, that the test concerns actual rather than ideal journals.

als' editorial boards and referees in rejecting a submission. But this evidence would by its nature be limited, since it would consist solely of the judgements of scientists acting as journal editors and referees. A better and more comprehensive investigation would not restrict itself to considering scientists' judgements when donning a particular kind of professional hat. The publication test, in sum, does not vindicate scientific platonism.[28]

7 *General principles of scientific method*

Scientific standards, according to the indifference objection, are indifferent between platonism and a non-platonist interpretation of mathematics equivalent to it. The platonist could respond that, despite their mathematical equivalence, standard scientific principles recommend the former. Take Hellman's modal-structuralism as an example of the latter. Modal-structuralism is a sophisticated, one might even say abstruse, interpretation of mathematics. Platonism is simpler and more familiar. Given that simplicity and familiarity are scientific principles, surely scientific standards recommend platonism over modal-structuralism? A second response to the indifference objection is thus that general scientific principles recommend platonism over any non-platonist interpretation of mathematics.[29]

Let us agree, as noted earlier, that some version of simplicity is a scientific virtue. It would be naive, however, to think that this means that scientific standards recommend any theory T_1 over any theory T_2 less simple than it. Suppose that T_1 and T_2 are spatiotemporally equivalent. For example, take T_1 to be the theory that there exists an omni-benevolent deity who has no effects on the spatiotemporal world, and T_2 the theory that a thousand such impersonal deities exist, all with different, highly complex moral characters. Both theories are consistent and by hypothesis have the same spatiotemporal import, namely: none! Clearly, however, T_1 is simpler than T_2. Does this mean that scientific standards recommend T_1 over T_2? It seems not—the apparent verdict is rather that science does not speak to the issue of which of T_1 and T_2 is true. Another example is the venerable one of theories with different units of measurement. A scientific theory based on the metric system is simpler than one based on British imperial units (feet, inches, etc.); a theory based on the Kelvin scale is similarly simpler than one based on that of Celsius; and so on. But superiority of this kind evidently does not (epistemically) privilege a theory over its more cumbersome counterpart.

These and countless other examples illustrate the point that the most general form of the principle of simplicity outstrips its scientifically accepted form. The precept 'Prefer *any* simpler theory to any more complex theory (in this respect)'

[28] A slightly more promising version of the publication (or publication-and-reception) test would invoke general scientific journals such as *Nature* or *Science* rather than *Physical Review*, since the former represent what scientists regard as important contributions to science in general rather than physics in particular. The same moral applies to these too, however.

[29] See e.g. Colyvan (2001: 128–9).

is not a principle of scientific theory choice. As the stress on 'any' shows, this formulation is too general: the real scientific principle has a more restricted range of application. The scientific platonist must therefore show that the appropriate restriction does not disable the principle from adjudicating between S^P and any acceptable S^{NP}.[30]

So far, our defence rests on the implausibility of thinking that a scientific principle such as simplicity holds with unrestricted generality, and the lesser, but still considerable, implausibility of thinking that a properly restricted version of the principle privileges S^P over all (acceptable) S^{NP}. It is worth backing up these impressions, well supported though they are, by turning to the specialist literature on the subject. Simplicity might be thought to be a shop-soiled theme in the philosophy of science, but some definite conclusions concerning it have emerged.

In a series of articles,[31] Elliott Sober has examined whether scientific uses of the principle of simplicity underpin its philosophical uses. More precisely, Sober has investigated the following question: Does the rationale for scientific uses of simplicity to adjudicate between predictively non-equivalent theories carry over to the case of predictively equivalent theories? Sober's considered answer is 'no'. For example, he argues that a leading model of how to measure the simplicity of solutions to the curve-fitting problem (which curve should be drawn through some given data points) does not apply to the decision between predictively equivalent philosophical theories (Sober 1996). On this model, due to the statistician Hirotugu Akaike, differences in the predictive accuracy of two families of curves arise from differences in how well they fit the data points together with a difference in the number of adjustable parameters they contain. For example, given some collinear data points, the predictive accuracy of the family of straight lines (parameterized by $y = ax + b$) is greater than that of the family of parabolic curves (parameterized by $y = ax^2 + bx + c$) because the goodness of fit of the closest-fitting member of each family is the same—namely, perfect—but the first family contains one less adjustable parameter than the second (a, b as opposed to a, b, c). As with some of the other examples Sober mentions, however, there is no difference in adjustable parameters between the spatiotemporally equivalent theories S^P and S^{NP}. Thus the simplicity criterion cannot take root here. Sober's measured conclusion (about theories that are predictively, not even spatiotemporally, equivalent) expresses the general point nicely.

> This treatment of the role of simplicity considerations in the curve-fitting problem provides no rationale whatever for choosing between theories that are predictively equivalent. This doesn't decisively *prove* that simplicity differences count for nothing in the case of predictively equivalent theories. However, it does lend support to that epistemological

[30] One might respond that the theories in my examples do not genuinely differ in their simplicity, and that *genuinely* simpler theories would be scientifically preferable to more complex ones. This response only differs terminologically from the position I am considering here, since the challenge would then be to demonstrate that a criterion of *genuine* scientific simplicity privileges S^P over any acceptable S^{NP}.

[31] In particular Sober (1996) and Forster and Sober (1994), as well as Sober (1994*a*; 1994*c*).

conclusion. To use the principle of simplicity in one context because it makes good sense in the other is to commit an epistemological equivocation (1996: 170).

Other writers have stressed that what grounds the use of simplicity in its scientific applications are particular features of the situation in question: remove this grounding and you thereby remove the scientific rationale for deploying the principle. Wesley Salmon for instance writes:

The most reasonable way to look at simplicity, I think, is to regard it as a highly relevant characteristic, but one whose applicability varies from one scientific context to another. Specialists in any given branch of science make judgments about the degree of simplicity or complexity that is appropriate to the context at hand, and they do so on the basis of extensive experience in that particular area of scientific investigation. Since there is no precise measure of simplicity as applied to scientific hypotheses and theories, scientists must use their judgement concerning the degree of simplicity that is desirable in the given context. The kind of judgement to which I refer is not spooky; it is the kind of judgement that arises on the basis of training and experience. This experience is far too rich to be the sort of thing that can be spelled out explicitly (1996: 279).

This passage makes it clear just how context-sensitive a truly scientific principle of simplicity is. It also indirectly supports the contention that any such application of simplicity to the case of S^{P} versus S^{NP} would be ungrounded, an inappropriate extrapolation beyond the contexts in which the principle finds its home. What one sees in the specialist literature more generally is suspicion of the idea that there is a universal, context-free scientific principle of simplicity. There is correspondingly little support for the idea that such a principle could provide scientific reasons for adjudicating between two spatiotemporally equivalent theories. In sum, methodological studies of the use of simplicity in science bid fair to rule out its use as a tie-breaker in the contest between S^{P} and S^{NP}. At the very least, uses of simplicity in standard scientific contexts are so different from its uses as a tiebreaker in philosophical controversies that the platonist's contentious extrapolation requires far more support than it has hitherto received. As we might put it: it is simplistic to identify scientific simplicity with simplicity *simpliciter*.

The moral applies to all scientific principles, not just simplicity. For example, it is often argued that even if a theory T_1 is in some sense (logically, spatiotemporally, etc.) equivalent to another theory T_2, T_1 may nevertheless be preferable to T_2 on account of its greater scientific fertility. A theory's fertility consists in how many new theories, new extensions of old theories, or new connections, results, methods, etc., it leads to. In short, T_1 may be 'statically' equivalent to T_2 but 'dynamically' superior to it, hence scientifically preferable. So if platonism were scientifically more fertile than anti-platonism, that would be a reason for preferring it.[32]

[32]This argument has been put forward by several philosophers, perhaps most fully in Baker (2001). See also Colyvan (2001: ch. 4), Steiner (2005: 644–5), and Quine's discussion of the importance of mathematical theories as 'engines of discovery' (1960: 270).

An immediate response to this argument is that there seems to be no difference in fertility between a platonist scientific theory T^{P} and its (acceptable) non-platonist counterparts T^{NP}. On a typical structuralist view, for example, actual mathematical usage can happily proceed with dummy variables, rendering it notationally indistinguishable from mathematical practice based on a platonist interpretation. If one has the structuralist interpretation in mind, apparent singular terms are understood as placeholders for positions in structures; if one has the platonist interpretation in mind, they are understood as bona fide singular terms. Everything else—every definition, manipulation, proof, etc.—goes through just the same. In the heat of mathematical practice, when background interpretation effectively drops out, there is no difference between the two. Alan Baker, who is a proponent of the argument, suggests that 'the application of group theory to particle physics ... allowed the prediction of whole families of hitherto unobserved subatomic particles' (2001: 92). But a structuralist interpretation of mathematics, with the everyday syntactic notation unchanged, would be just as capable of allowing this prediction as a platonist one. Likewise for the development of quaternions, and any other example one might care to mention. Differences in background mathematical interpretation do not seem to affect a theory's long-term scientific fertility.

Another response is also available in this case, which parallels the response in the case of simplicity and which is less dependent on the contingencies of humans' adeptness with various mathematically equivalent theories. Even if T^{P} were easier to handle and therefore ultimately more generative of scientific developments than T^{NP}, the two theories would not necessarily differ in their scientific fertility properly so-called. Taking modal structuralism as an example once more, we may confidently predict that all the mathematics that will ever be needed by science is contained within ZFC, which modal structuralism can accommodate (as we are assuming). Any proofs, results, techniques, etc., to be found in platonist mathematics that could conceivably find scientific application have modal-structural analogues. And the spatio-temporal equivalence of platonism and modal structuralism is true not just for the static snapshot of current scientific practice, but for any conceivable development thereof. Perhaps platonist mathematics is slightly easier for the human mind to grasp, and therefore takes up fewer intellectual resources, freeing scientists to spend those extra resources on scientific inquiry. But that is not in itself a scientifically relevant difference in fertility. The case of theories with different units of measurement again serves as an example here.

In short, mere psychological suggestiveness does not a scientifically fertile theory make; at least not in the sense relevant to scientific confirmation. The boundary between the context of justification and the context of discovery may be more blurred than has often been assumed, but a pertinent distinction of this kind must obtain, on pain of making even the most arbitrary features of scientific theories and the most idiosyncratic aptitudes of individuals who deploy them relevant to scientific confirmation. Anything can suggest anything to anyone. What counts is what avenues the theory opens up in some appropriately logical sense, not the long-term production rate of the scientific community that adopts it. By assumption, every scientific application or implication of platonistic theory T^P is one that is shared by any of its acceptable non-platonist equivalents T^{NP}. Of course, one difference between the imperial-metric units case and that of a platonist versus a non-platonist version of the same scientific theory is that the latter contrast is between two theories with different contents, whereas by assumption the imperial-unit and metric-unit versions of a scientific theory share the same content. But my point is precisely that various such differences must be shown to be of scientific relevance. So the question is whether the potential advantage in simplicity possessed by platonism, as a result of which it may—possibly—in the long run give rise to more developments than its non-platonist counterpart is relevant to its scientific superiority and confirmation.

These general difficulties for scientific platonism have been obscured, I believe, by an overemphasis on scientific principles at the expense of their context of application. Wrenching principles such as simplicity or fertility out of their proper domain is a snare against which every philosopher must guard. Scientific standards recommend the application of scientific principles and methods to certain kinds of contexts and questions. It is thus not enough to show that the application of general scientific principles to the choice of S^P versus some S^{NP} settles the issue one way or the other; one must also show that the question is scientific. The opposed view is that the issue is beyond the scientific pale, that to apply these principles to this particular question is to transcend the scientific realm. The debate must be decided not by table-thumping declarations from either side that the question is or is not scientific, but by examining the exact boundary science itself posits between questions within the scientific realm and those outside it. As explained, the prospects for platonism on this point are not promising.

One reaction by the scientific platonist might be to try to reformulate her claim in terms of generalized scientific principles rather than scientific grounds. The revised claim would be that generalized scientific principles (be they applied to a scientific question or not) endorse platonist mathematics. A generalized scientific principle is one with the restrictions demanded by proper scientific method removed, as in the most general version of simplicity (roughly, the non-scientific 'prefer the simpler of *any* two theories' as opposed to the scientific 'for any two theories of such-and-such kind prefer the simpler of the two'). This revised claim, however, would be unfaithful to the inspiration behind the scientific platonist position, which is to take scientific grounds—and not generalized scientific principles extrapolated beyond the domain of science—as authoritative. What lends the scientific platonist thesis its interest is the importance attached to *scientific* standards as opposed to ultra-scientific generalizations of these standards. Given any metaphysics of mathematics, it is of course a trivial exercise to concoct a set of principles that support it over its rivals; the question is why anybody should be interested in what *those* particular principles support. Anything gained by this reformulation therefore seems to be divested of the force it was intended to have.

It is in effect to give up on scientific platonism.

The scientific platonist might respond in a different way, by expanding the standard conception of the sciences to include the non-natural sciences. For example, one might classify semantics as a science and advocate semantic uniformity as a reason for preferring platonism to non-platonism. After all, a realist semantics for mathematics is in line with the proper semantics for (most of) the rest of language (a point famously stressed in Benacerraf (1973)). Mathematical and non-mathematical language *appear* very similar syntactically and semantico-inferentially: noun phrases, predications, connectives, quantifiers, etc., all seem to function in the same way. The presumption of uniformity applies with particular force to so-called mixed statements containing a mixture of mathematical and empirical language, which are usually read in a semantically seamless way: a stock example is, 'If the first ball bearing has mass m_1 and the second ball bearing has mass m_2 then the gravitational force between the two masses at distance d is Gm_1m_2/d^2.' So perhaps it will be said that our default semantics should take this apparent similarity to result from a real similarity, and hence that scientific standards favour realism over other interpretations of mathematics.

This semantic argument is undoubtedly to be taken seriously. One missing step is of course the transition from realism to platonism. But a more pertinent problem with the argument in the present context is that to resort to it is simply to abandon scientific platonism. Anyone who concedes that natural-scientific grounds do not support platonism but holds out hope that semantic grounds support platonism has given up on scientific platonism and adopted semantic platonism instead (as by analogy we might label the view). Lest the reader should think that I have deliberately attacked the weaker thesis and dodged the really powerful argument, however, I should briefly sketch some reasons for thinking that the argument for scientific platonism is more likely to succeed than the argument for semantic platonism.

First, semantics is a relatively young discipline with many competing research programmes. In particular, there is nothing like agreement across the board that the realist view of mathematical language is correct. Second, many philosophers would incline to the view that the correct semantic theory for mathematics must flow from a metaphysics for mathematics rather than the other way round. Indeed they would argue that an adequate semantics for mathematics can always be fashioned from any reasonable metaphysics. On this view, metaphysics is the horse and semantics the cart,[33] and tackling the semantics of mathematics

[33] As Michael Jubien, who vividly formulates the view (about properties), has put it, 'It is easy to think we must look hard at natural language semantics in order to evolve a theory of properties. Indeed, it is easy to think even that property theory is nothing *but* a certain region of natural language semantics. But once we reject conceptualism in favour of genuine realism, matters seem quite different. It then looks like a theory of properties should flow directly from general *metaphysical* considerations, and that the *semantical* projects should be founded upon the results of the metaphysical one. Of course none of this is to say that our metaphysical intuitions occur in some vacuum, independently of language, or that linguistic *data* are not influential in the formation of our metaphysical views. It is only to say which theoretical enterprise is the cart and which the horse. In my view, metaphysics is the horse. Which is not to disparage the cart at all, but only to suggest that it can't go very far on its own,

without having its ontology before us is a methodological solecism. Of course we can agree that trying to settle the metaphysics in a state of semantic naivety is a recipe for disaster; but this does not imply that the latter should take precedence over the former. I stress that I am not peddling this methodological view myself; I am merely pointing out its respectability. Finally, and perhaps most importantly, semantic platonism's significance, philosophically speaking, is also thought to be less than that of scientific platonism, the reason being that what semantic standards support is generally thought to have less of a claim on us than what scientific standards support (this is in turn related to the first claim). Steven Wagner, who is more sympathetic to semantic platonism than most, once again puts the point crisply.

> The problem [with semantic platonism compared to scientific platonism] is that science has clearer credentials than formal semantics. Physics is acceptable beyond doubt. If it admits no nominalistic construal, then Platonism is true. Semantics, however, is not clearly science and not clearly anything else that compels belief in its ontology. (1996: 77)

So I leave the examination of semantic platonism for another occasion, and trust that the reader will appreciate that this is not tantamount to discussing *Hamlet* without mentioning the prince, even if it is not quite tantamount to discussing *Hamlet* without mentioning the second gravedigger.[34]

Observe finally that the scientific platonist must also contend with an intermediate position, according to which it is indeterminate whether scientific standards favour S^{P} over (any acceptable) S^{NP}, and perhaps more generally whether they ever favour one of two spatiotemporally equivalent theories. On this view, there is no fact of the matter. Perhaps the grounds for, say, positing an entirely general principle of simplicity as governing scientific practice may be very strong. The grounds for resisting its extrapolation to the non-spatiotemporal, however, may be just as strong. In other words, the practice may be indeterminate as to whether S^{P} is scientifically preferable to S^{NP}. The platonist must also defeat this indeterminacy claim if she is to uphold her position, since it represents a determinate stand.

8 *Scientific grounds and the actual content of mathematics*

According to the indifference objection, scientific grounds endorse mathematics but not its platonist interpretation. For example, scientific grounds endorse the truth of '$2 + 3 = 5$' under some acceptable interpretation but they do not endorse its platonic interpretation 'the abstract numbers 2, 3 and 5 stand in the abstract relation of addition'. A statement such as '$2 + 3 = 5$' must, however, have *some* interpretation in its standard mathematico-scientific contexts. (Formalism, which

especially if it thinks the horse is its cargo.' (1989: 164).

[34]For further discussion of the 'semantic indispensability argument', see Colyvan (2001: 15–17) and especially Wagner (1996), where, having noted the quoted difficulty for semantic platonism, the author ends up advocating it anyway (or at least 'bets on it').

takes mathematics as actually practised to be meaningless symbol-manipulation, is wrong.) This observation gives rise to an objection and a challenge. The objection is that the standard content of mathematical sentences is realist (or perhaps even platonist), and therefore, since scientific grounds endorse this standard content, they endorse realism (or platonism). (This is essentially the argument we met in the first paragraph of section 3.) But even if the standard content is not realist (nor platonist), a challenge remains: What exactly *is* it that scientific grounds endorse?

Let me explain why the objection is misconceived before saying a few words about the challenge. The objection assumes that scientific grounds endorse the standard content of mathematics. But there is no good reason to think that scientific grounds and our best semantic theory are aligned. Indeed, if the assumption that the content of standard mathematical utterances (and inscriptions, etc.) is realist is correct, then as a matter of fact the two are *not* aligned. The claim that the realist interpretation captures the content of standard mathematics is thus consistent with the claim that scientific standards do not endorse that interpretation over, say, a structuralist one.

Why would anyone think otherwise? Let us return to '$2 + 3 = 5$' to fix ideas. The argument applied to this example runs as follows: (i) our best semantic theory informs us that the content of '$2 + 3 = 5$' in standard scientific contexts is realist; (ii) scientists (*qua* scientists) endorse the proposition expressed by standard utterances of '$2 + 3 = 5$'; (iii) scientists' endorsement (*qua* scientists) is our best guide to what scientific standards endorse; hence scientific standards endorse the realist interpretation of '$2 + 3 = 5$'. Note that this argument does not assume that scientists have privileged access to their utterances' content. The argument is not deductively valid of course; but if its premises were all true it would take special pleading to attribute widespread error to the scientific community. Premise (iii) seems unimpeachable, and we are also granting (i) for the sake of argument. That leaves premise (ii) as the potential site of dispute.

There seems to be independent reason, however, for rejecting premise (ii). The proponent of the indifference objection will urge that scientists (*qua* scientists) endorse the claim that some (acceptable) interpretation of '$2 + 3 = 5$' is true but not that it is true *tout court*. Of course, scientists casually say that '$2 + 3 = 5$' is true, but this seems to be unreflective usage, abbreviating a more complex truth-endorsement. In more reflective moments, scientists will recognize that what they endorse (*qua* scientists) is rather, as we have put it, that the sentence '$2 + 3 = 5$' is true on some acceptable interpretation. This point is independently verifiable—though of course scientists might not put it in these very terms. Casual acquiescence in sentences that a semantic theory construes realistically should not be mistaken for reflective endorsement of the statement on the interpretation that semantic theory gives it (or any other interpretation). But what is certainly not casual is the truth-endorsement.

The point is perhaps best put in the following more general way. To the extent that scientists recognize that some specific '...ism' (if any) is the correct interpretation of their mathematical discourse, to that extent they do not endorse their utterances under that interpretation (*qua* scientists). Thus scientists' casual willingness to accept whatever is the content of their mathematical utterances is trumped by their recognition that scientific standards do not endorse realism any more than they endorse a non-realist interpretation equivalent to it.[35] In sum, even if realism (or platonism) turns out to be the proper interpretation to put on standard utterances, it does not follow that this is the scientifically endorsed interpretation. In fact, if successful, what the arguments in the previous sections show is precisely that it is *not* scientifically endorsed.

A belt and braces response to the objection could go on to argue that structuralism might be a better candidate, all things considered, for the standard content of mathematical utterances, at least when it comes to the non-set-theoretic mathematics contained in scientific applications (this would be to challenge premise (i)). A yet further response would be to ask how the transition from realism to platonism is supposed to be effected. Alas there is no room to examine either of these further responses here, so let me say a quick word about the challenge. I shall indicate the first step along a path that the anti-platonist will wish to explore further.

The proponent of the indifference objection owes us a model of how scientific standards can endorse the truth of '$2 + 3 = 5$' without endorsing any particular content. A simple model might be that scientific standards endorse a disjunction of acceptable interpretations of the sentence '$2 + 3 = 5$' without endorsing any particular one.[36] Another model, which we have in fact been presupposing throughout, is that scientific standards endorse the proposition that the sentence '$2 + 3 = 5$' is true under some acceptable interpretation but not any particular one. Note that this latter commitment is existential and metalinguistic. The two proposals are not contradictory, of course, and in fact form a natural package: should scientific standards endorse the disjunction as well as the existential claim, both proposals would be correct. Observe that on the disjunctive model, the propositional operator 'scientific standards endorse ... ' applies to a disjunction but to none of its disjuncts, and that on the existential model the same operator applies to an existentially quantified statement but to none of its instantiations. This logical behaviour is familiar from the properties of epistemic operators in general. Many such operators apply to disjunctions without applying to any disjunct: for example it may be known, reasonably believed, thought, etc., that one of p or q is true but not that p is true or that q is true. Similarly, such operators can apply to existentially quantified statements without applying to any of their specific instances: for example, I know that there is some truth or other that I do not know, but there is no specific truth p such that I know that: p is true but I do not know p; or I know that some combination opens the lock but not which one;

[35]Of course a particular scientist might have her own individual non-scientific reasons in favour of some particular interpretation.

[36]The disjunction consists of two or more disjuncts.

and so on. The epistemic operator 'scientific standards endorse ... ' as applied to mathematical interpretations seems to be another instance of this phenomenon.

Clearly, these models—better, templates for a model—are in need of further elaboration, which I cannot give them here. But my defence to this section's objection is independent of this elaboration, as it rests on the earlier point about the relation between scientific grounds and semantic theory. What remains for the anti-platonist to do is to explore possible models of how scientific grounds might endorse the claim that a mathematical statement is true under some interpretation, or some disjunction of interpretations, etc., without endorsing any particular interpretation. If the earlier arguments are sound, some such model must be correct.

9 *Conclusion*

Our discussion has thrown up a challenge for scientific platonism. Even if the scientific platonist can overcome the pragmatic objection—that scientific standards do not endorse the truth of mathematics but only acceptance of it in some non-epistemic sense—she must still confront the indifference objection, that scientific standards endorse that mathematics is true under some interpretation or other but not under any particular one such as platonism or realism or structuralism or modal-structuralism or .... The indifference objection defeats strong scientific platonism, according to which scientific standards endorse the platonist interpretation of (at least some parts) of mathematics. Whether it defeats weak scientific platonism, according to which scientific standards endorse the thesis that mathematics is committed somewhere down the line to abstract objects, depends on whether any acceptable S^{NP} must ultimately be committed to abstract objects. That question, as explained, remains open.

Sections 6–8 considered three responses to the indifference objection on behalf of the scientific platonist. The first is that letting scientific referees decide the issue, by the usual means and in the usual fora, vindicates platonism. The second is that general scientific principles recommend platonism over non-platonism. The third is that scientific practice itself sanctions whatever is the standard—namely, realist (or perhaps platonist)—interpretation of mathematical statements. I take my responses to these objections to have shown that the scientific platonist cannot gain a quick victory by any of these means. On the question of science's relevance to the metaphysics of mathematics, the typical mathematician may be right after all.[37]

[37]I have benefited from presenting this material at the universities of York, Oxford, and Wisconsin-Madison as well as at the Cambridge conference. My thanks also to two anonymous referees for this volume and to the following for discussion or comments at various stages: Alex Oliver, David Liggins, Elliott Sober, Gideon Rosen, Hallvard Lillehammer, Mary Leng, Michael Potter, Paul Benacerraf, Peter Smith, Sarah Teichmann, Tim Lewens, and Tim Williamson.

On Quantifying into Predicate Position: Steps towards a new(tralist) perspective

CRISPIN WRIGHT

In the *Begriffsschrift* Frege drew no distinction—or anyway signalled no importance to the distinction—between quantifying into positions occupied by what he called *eigennamen*—singular terms—in a sentence and quantification into predicate position or, more generally, quantification into open sentences—into what remains of a sentence when one or more occurrences of singular terms are removed. He seems to have conceived of both alike as perfectly legitimate forms of generalization, each properly belonging to logic. More accurately: he seems to have conceived of quantification *as such* as an operation of pure logic, and in effect to have drawn no distinction between first-order, second-order and higher-order quantification in general.

In the twentieth century the prevailing view, largely under the influence of writings of and on Quine, became quite different. The dominant assumption was of the conception of quantification encapsulated in Quine's famous dictum that 'to be is to be the value of a variable'. The dictum, as we know, wasn't really about existence but rather about ontological *commitment*. Subsequently, it was better expressed as 'to be *said to be* is to be the value of a variable'. Writing in a context when ontic fastidiousness was more fashionable than now, Quine was preoccupied with possibilities of masked ontological commitments and, conversely, of merely apparent ontological commitments. His proposal was: regiment the theory using the syntax of individuation, predication, and quantification and then see what entities you need to regard as lying in the range of the bound variables of the theory if it is to rank as true. You are committed, as a theorist, to what you need to quantify over in so formulating your theory. (As to the prior question of how we are supposed to recognize the *adequacy*—or inadequacy—of such a regimentation, Quine was largely modestly silent.)

The Quinean proposal can seem almost platitudinous. But once it is accepted, first- and second-order quantification suddenly emerge as standing on a very different footing. First-order quantification quantifies over *objects*. No one seriously doubts the existence of objects. By contrast, second-order quantification seems to demand a realm of *universals*, or *properties*, or *concepts*. And of such entities Quine canvassed an influential mistrust: a mistrust based, initially, on their mere abstractness—though Quine himself later, under pressure of the apparent needs of science, overcame his phobia of the abstract—but also on the ground that they seem to lack clear criteria of identity—a clear basis on which they may be identified and distinguished among themselves. It was the latter consideration which first led Quine to propose that the range of the variables in higher-order logic might as well be taken to be *sets*—abstract identities no doubt, but ones with a clear criterion of identity given by the axiom of extensionality—and then eventually to slide into a view in which 'second-order logic' became, in effect, a misnomer—unless, at any rate, one regards set theory as logic.[1] By 1970 he had come to his well-known view:

> Followers of Hilbert have continued to quantify predicate letters, obtaining what they call higher-order predicate calculus. The values of these variables are in effect sets; and this way of presenting set theory gives it a deceptive resemblance to logic. ... set theory's staggering existential assumptions are cunningly hidden now in the tacit shift from schematic predicate letters to quantifiable set variables (Quine 1970: 68).

Those remarks occur in Quine's paper 'The Scope of Logic' in the sub-section famously entitled: Set Theory in Sheep's Clothing! By the end of that paper, Quine has persuaded himself, and probably most of his readers too, that Frege and others such as Russell and Hilbert who followed him in allowing higher-order quantification have simply muddied the distinction between logic properly so regarded—the theory of the valid patterns of inference sustained by the formal characteristics of thoughts expressible using singular reference, predication, quantification and identity—and set theory which, to the contrary, is properly regarded as a branch of mathematics.

My principal concern in what follows will be to outline what I think is the proper, and overdue, reaction to—and against—the conception of quantification which drives so much of the Quinean philosophy of logic: the conception which sees quantification into a particular kind of syntactic position as essentially involving a *range of entities* associated with expressions fitted to occupy that kind of position and as providing the resources to generalize about such entities. This view of the matter is so widespread that it has become explanatory orthodoxy. For instance in the entry under 'Quantifier' in his splendid *Oxford Dictionary of Philosophy*, Simon Blackburn writes that 'informally, a quantifier is an expression that reports a quantity of times that a predicate is satisfied in some class of *things*, i.e. in a "domain"' (1994: 313, my emphasis), while the corresponding entry in Anthony Flew' and Stephen Priest's *Dictionary of Philosophy* observes that 'the truth or falsity of a quantified statement ... cannot be assessed unless one knows what totality of objects is under discussion, or where the values of the variables may come from' (2002: 338). I believe that this conception—of quantification as essentially *range-of-associated-entities-invoking*—is at best optional and restrictive and at worst a serious misunderstanding of what quantification fundamentally is; more, that it squanders an insight into the nature of the conceptual resources properly regarded as logical which Frege already had in place at the time of *Begriffsschrift*.

[1] The Quinean tradition has, of course, absolutely no regard for Frege's notion of a *logical object*.

Why is this an issue apt for discussion in an anthology whose focus is the epistemology of mathematics? What has the question of the proper conception of higher-order quantification got to do with mathematical knowledge? Part of the answer lies in the integral role played by higher-order logic in the contemporary programme in the philosophy of mathematics—sometimes called neo-Fregeanism, or neo-logicism but I prefer *abstractionism*—which seeks to save as much as possible of the doomed project of Frege's *Grundgesetze* by replacing his inconsistent Basic Law V with a number of more local *abstraction principles* designed to provide the deductive resources for derivations of the fundamental laws of, for example, classical number theory and analysis in systems of second-order logic.[2] There has been much discussion of the status and credentials of abstraction principles for their part in this project. But one thing is clear: if the *logic* used in the abstractionist programme is indeed, as Quine thought, nothing but set theory in disguise, then execution of the various local abstractionist projects, however technically successful, will be of greatly diminished philosophical interest. A reduction of arithmetic, or analysis, to a special kind of axiom cum *set theory* will hardly be news! It is therefore, generally speaking, a presupposition of the significance of the abstractionist programme that there be a quite different way of thinking about higher-order logic than the Quinean way—and that this different way, whatever exactly it is, should be consonant with the general spirit of logicism: the thesis that logical knowledge and at least basic mathematical knowledge are, in some important sense, of a single epistemological kind.[3]

The connection with the abstractionist programme is, however, only part of the reason why the epistemologist of mathematics should interest herself in the nature of higher-order logic. It is probably fair to say that contemporary philosophers of mathematics number many more structuralists than neo-logicists, and among structuralists there is an increasing trend of placing weight on higher-order logics both as sources of categorical axioms—presumed to provide determinate characterizations of the structures of which pure mathematics is conceived as the special science—and as a safe medium for their investigation. For this programme, too, pure mathematical knowledge is heavily invested in and conditional on the good standing of the concepts and deductive machinery of higher-order logic.

To develop a new overarching conception of quantification which is powerful enough to make sense of the classical logic of the *Begriffsschrift* yet stands contrasted with the Quinean orthodoxy in the way prefigured is a tall order. The task is part philosophical and part technical. My goal in what follows has to be limited. I shall attempt no more than to provide some primary motivation for the thought that such an alternative—non set-theoretic, properly logical—conception

[2]A brief sketch of the current situation with respect to the prospects of technical recovery of arithmetic, analysis, and set theory on an abstractionist basis is provided in the Appendix.

[3]The kind in question is sometimes gestured at by the inflammatory term, 'analytic'. A more useful characterization will appeal to a class of concepts whose content allows of explanation by means that (ancestrally) presuppose a familiarity only with logical concepts, and a class of truths formulable using just such concepts which can be recognized to be true purely on the basis of those explanations and logical deduction.

does indeed exist, and to outline some of the cruces which will have to be negotiated in developing it properly. In short, my aim here is no more than to outline a project, as a spur to further work.

1 *Basic idea and project outline*

The proposal is to explain and vindicate what we may term a *neutralist*[4] conception of higher-order quantification—and indeed of quantification more generally. Here is a quotation from a recent paper by Agustín Rayo and Stephen Yablo that I think gives us a pointer in the right direction:

> If predicates and the like needn't name to be meaningful—to make their characteristic contribution to truth-value—then we have no reason to regard them as presupposing entities at all. And this indeed appears to be Quine's view. But now he goes on to say something puzzling:
>
> One may admit that there are red houses, roses, and sunsets, but deny, except as a popular and misleading manner of speaking, that they have anything in common. (Quine 1953: 10)
>
> Quine is right, let's agree, that 'there are red houses, roses and sunsets' is not committed to anything beyond the houses, roses and sunsets, and that one cannot infer that 'there is a property of redness that they all share.' But why should 'they have something in common'—or better, 'there is something that they all are'—be seen as therefore misleading? If predicates are non-committal, one might think, the quantifiers *binding* predicative positions are not committal either. After all, the commitments of a quantified claim are supposed to line up with those of its substitution instances. Existential generalizations are *less* (or no more) committal than their instances, and universal generalizations are *more* (or no less) committal. 'There is something that roses and sunsets are' is an existential generalization with 'roses and sunsets are red' as a substitution instance. So the first sentence is no more committal than the second. But the second is not committed to anything but roses and sunsets. So the first isn't committed to anything but roses and sunsets either (Rayo and Yablo 2001).[5]

There is a principle implicit in this passage that I want to endorse. Here is a formulation:

> (*Neutrality*) Quantification into the position occupied by a particular type of syntactic constituent in a statement of a particular form cannot generate ontological commitment to a kind of item not *already* semantically associated with the occurrence of that type of constituent in a true statement of that form.

This is less than the whole gist of the quoted passage, since Rayo and Yablo are also claiming—in a nominalist spirit—that the ontological commitments of 'Roses and sunsets are red' go no further than roses and sunsets. What I am endorsing,

[4]This apt term was Fraser MacBride's suggestion

[5]A similar thought is expressed in Hale and Wright (2001: 431–2, Postscript, problem #11). The basic idea is anticipated at Wright (1983: 133).

by contrast, is a more modest claim: precisely, that quantification is *neutral* as far as ontological commitment is concerned—that the commitments of quantified statements go no further than the requirements of the truth of their instances *whatever the latter requirements are*. So the view I am suggesting is also open to someone who thinks that simple predications of 'red' *do* commit us to, say, a universal of redness, or to some other kind of entity distinctively associated with that predicate. But the crucial point is that Quine is not on board in any case. For in Quine's view, while mere predication *is* free of distinctive ontological commitment, quantification into predicate position must commit a thinker to an *extra* species of entity—in the best case, sets; in the worst, attributes or universals. This, I think, is a major mistake.

The basic idea I wish to set against Quine's is that quantification should be viewed as a device for *generalization of semantic role*. Given any syntactic category of which an instance, *s*, can significantly occur in a context of the form [...*s*...], quantification through the place occupied by '*s*' is to be thought of as a function which takes us from [...*s*...], conceived purely as a content, to another content whose truth-conditions are given as satisfied just by a certain kind (and quantity) of distribution of truth values among contents of the original kind. A quantifier is a function which takes us from a statement of a particular form to another statement which is true just in case some range of statements—a range whose extent is fixed by the quantifier in question—which share that same form are true. The central task for the project I am proposing is to sharpen and develop this rather inchoate idea.

Four issues loom large on the agenda. The primary task will be

> To develop a definite proposal, or proposals, about how quantification, neutrally conceived, can work.

There are then three further interrelated tasks. The second is

> To explore whether neutralism can provide a satisfactory account of—that is, sustain all the demands on quantification made by—classical impredicative higher-order logics.

Should matters stand otherwise, the third task will be

> To explore what kinds of higher-order logic *can* be underwritten by neutralism.

The fourth task will be

> To determine whether neutralism can sustain the epistemological and technical demands imposed on higher-order logic by the abstractionist programme, and by other programmes in the foundations of mathematics which make special demands on, or have special needs for, higher-order theories.

There are, of course, other potentially 'ontologically liberating' conceptions of higher-order quantification. Notable among them are substitutional construals and Boolos's conception of such quantifiers as in effect devices of plural generalization, ranging over the same items as first-order quantifiers but ranging over them in bunches, as it were.[6] The shortcomings of these approaches are well known: Boolos has no ready means of interpreting quantification over relations, and substitutionalism limits legitimate generalization to the range of cases captured by the expressive resources of a particular language. Neutralism must do better. It must provide a uniform construal of higher-order generalization, irrespective of the grade of predication involved; and it must allow language-unbounded generalization. Well executed, it will not of itself put to bed Quine's 'set-theory in sheep's clothing' canard. But it will do so in conjunction with the thesis that a satisfactory account of predication need involve no commitment to sets.

2 *Fixing the meanings of the quantifiers*

The obvious—I think inevitable—direction to explore is to suppose that the conception of quantification we seek must ultimately be given by direct characterization of the *inferential role* of the quantifiers; more specifically, that what is needed are explicit statements of natural deductive introduction- and elimination-rules for the various possible {syntactic-type : quantifier-type} pairings. In order to be effective—to be safe in inference and to establish genuine concepts—such stipulations must of course meet certain general constraints: at a minimum consistency, but also arguably conservativeness and harmony.[7] Some of the issues here are on the abstractionist agenda already, in connection with the task of characterizing which are the a priori acceptable abstraction principles.[8] In any case, the proposal will be to view suitable rules as *constituting meanings* for the quantifiers and—simultaneously—as showing how they can assume (something close to) their standard inferential roles without being saddled with the 'range-conception' of their function which drives the Quinean critique.

[6]See especially his (1984).

[7]Familiarly, these notions are open to various construals. Here I intend only the general idea that the patterns of inference, in and out, licensed by putatively meaning-constitutive rules should have a certain theoretical coherence: that what one is licensed to infer from a statement by such (elimination) rules should be exactly—no more and no less—what is prepared for by the (introduction) rule that justifies one in inferring to it. Non-conservativeness—for example, to the egregious degree illustrated by Arthur Pryor's connective 'tonk'—is the penalty of violating such constraints one way round. But disharmony in the converse direction—for example, that manifested by a connective, *tunk*, which combines the disjunction-elimination and conjunction-introduction rules—results in safe yet *unintelligible* inferential practice. Inferentialism is a broad church, and there are many ways of developing the basic thought that the most fundamental logical laws are constitutive of the logical operations that they configure, and that this consideration is somehow key when it comes to explaining our knowledge of them. I cannot pursue these issues here, nor attempt to respond to the recent opposition to the basic thought developed by writers such as Timothy Williamson in his (2003) and (2006). The aim of the present chapter must be restricted to a conditional: if inferentialism can contrive a satisfactory account of the epistemology of basic logic at all, then it can play this part in relation to higher-order quantification theory in particular, and in a way that can liberate our understanding of it from the Quinean shackles.

[8]See the discussion in (Hale and Wright 2000).

So: what might be the shape of suitable such rules? If quantification is to be conceived as an operation which generalizes on semantic role in a uniform way, irrespective of what the semantic role of the particular kind of expression concerned actually is, then it might seem to be a key thesis for neutralism that the various kinds of quantifiers—and *par excellence* the universal and existential quantifiers—each admit of a *single* account which goes through independently of the syntactic category of the expressions which the bound variables replace. And indeed I do not know that such an account is not attainable. However while natural deductive formulations of the quantifier rules for higher-order logic, for instance, should, of course, follow their first-order cousins as closely as possible, it is not clear that neutralism *demands* strict uniformity of explanation. It should be good enough merely if there is some suitable common pattern exhibited, for example, in the different kinds of universal quantification, even though the statements of the proof-rules for, say, first- and second-order universal quantification display important local differences and cannot be arrived at simply by applying a more general uniform account to the two syntactic classes of expression concerned. Consider for instance the standard first-order Universal Introduction rule:

$$\frac{\Gamma \Rightarrow A(t)}{\Gamma \Rightarrow (\forall x)Ax}$$

This provides that, if we can establish $A(t)$ on a certain pool of assumptions, Γ, then—provided the proof meets certain additional requirements—those same assumptions suffice for the universal quantification through 't'. The additional requirements in question are, of course, standardly characterized syntactically—for the kinds of formal language usually envisaged, it suffices to specify that the sentences in Γ be free of (unbound) occurrences of 't'—but their intent is to ensure that the proof in question has a certain *generality*: that the derivation of $A(t)$ in no way turns on the choice of 't' but would succeed no matter what (non-empty) term we were concerned with. That's the very feature, of course, that makes it legitimate to generalize. So in order for a proposed introduction rule for a second-order operator to count as governing a genuinely universal quantifier, the same feature must be ensured. Thus we would expect to retain the basic pattern

$$\frac{\Gamma \Rightarrow \Phi(F)}{\Gamma \Rightarrow (\forall X)\Phi X}$$

to which would then be added constraints to ensure that the derivability of $\Phi(F)$ from the set of assumptions in Γ in no way turns on the choice of the predicate F but would go through no matter what predicate was involved.

However it is clear that, save where F is atomic, the intended generality cannot be ensured merely by stipulating that Γ be free of occurrences of F. For F may be an open sentence of arbitrary complexity. If its satisfaction-conditions are logically equivalent to those of some syntactically distinct open sentence, F^*, then the derivation of $\Phi(F)$ from Γ may in effect lack the requisite generality—since Γ may incorporate special assumptions about F^*—even though the strict analogue of the first-order restriction is satisfied. Γ may just be the assumption $Fa \vee Ga$, for example. In that case, propositional logic will take us to $Ga \vee Fa$. And now, since '$Fa \vee Ga$' is free of any occurrence of the 'parametric' predicate, $G\ldots \vee F\ldots$, the envisaged second-order UI rule would license the conclusion, $Fa \vee Ga \Rightarrow (\forall X)Xa$.

Obviously we do not want to outlaw all generalization on semantically complex predicates. The way to ensure the generality we need is to require that the proof of the premise-sequent, $\Gamma \Rightarrow \Phi(F)$, be such that it will still succeed if the occurrence of 'F' in its conclusion is replaced by an atomic predicate. That way we ensure that any semantic or logical complexity in F is unexploited in the proof, which consequently is forced to have the requisite generality. A similar requirement on UI will indeed be apposite at first-order too if the language in question permits the formulation of complex—for instance, definite descriptive—singular terms. But my point is merely that the spirit of neutralism will not be violated if the eventual rigorous statement of the second-order rule proves to differ in a number of respects from its first-order counterpart: in order to ensure that both are concerned with universal quantification, properly so regarded, it will be enough that they exhibit the common structure illustrated and that their differences flow from the respectively different demands imposed by the need for generality, in the light of the differences between the syntactic types of expression concerned.

Semantic complexity in expressions quantified into raises another issue that it is worth pausing on briefly. In a language featuring definite descriptive singular terms, some form of free-logical restriction on the standard first-order rules may be required to ensure that the validity of the transitions in question is not hostage to reference failure. We might impose on the UE rule for instance

$$\frac{\Gamma \Rightarrow (\forall x)Ax}{\Gamma \Rightarrow A(t)}$$

the restriction that 't' may not be just any term but must be such that, for some atomic predicate, F, we have in addition that $F(t)$; so that a fuller statement of the rule would look like this:

$$\frac{\Gamma \Rightarrow (\forall x)Ax; \quad \Delta \Rightarrow Ft}{\Gamma, \Delta \Rightarrow A(t)}$$

The point, intuitively, would be that while an empty term may feature in true extensional contexts—for instance, 'there is no such thing as Pegasus'—its lack of reference will divest any *atomic* predication on it of truth. Now, the relationship between sense and reference for the case of predicates (if we permit ourselves for a moment to think of predicates as having reference at all) is manifestly different to that which obtains in the case of terms. There can be no presumption that a complex singular term refers just in virtue of its having a sense, whereas in the general run of cases the association of a predicate, complex or simple, with a determinate satisfaction-condition ensures it against (any analogue of) reference failure. But the point is not exceptionless: the predicate '... is John's favourite colour' as it occurs in

That car is John's favourite colour

may, despite being a significant open-sentence of English associated with determinate satisfaction-conditions, actually fail of reference—or anyway suffer an appropriately analogous failing to that of an empty but meaningful singular term—in just the same circumstances, and for the same reasons, as the singular term, 'The favourite colour of John'. A fully general natural deductive formulation of higher-order logic would have to allow for this kind of case. However it could not happily do so merely by transposing the reformulated first-order UE rule above, as

$$\frac{\Gamma \Rightarrow (\forall X)\Phi(X);\quad \Delta \Rightarrow Ft}{\Gamma, \Delta \Rightarrow \Phi(F)}$$

where 't' is required to be atomic, since that would be tantamount to the requirement that, before F could be legitimately regarded as an instance of a second-order universal quantification, we need it to be *instantiated*—for there to be something which is F. But that is not the right requirement: what we need is not that something be F but rather, putting it intuitively, that there be something which being F is. (By contrast, John might have a favourite colour that was unexemplified by anything.)

It needs further discussion how this wrinkle might best be addressed. It would be natural to propose that we might mimic the first-order restriction, but require that the collateral premise ascend a level, so that what is required is that F be correctly characterized by some atomic *second-order* predicate (in effect, a quantifier). Thus the statement of the rule would be

$$\frac{\Gamma \Rightarrow (\forall X)\Phi(X);\quad \Delta \Rightarrow \Psi F,\quad \text{`}\Psi\text{' atomic}}{\Gamma, \Delta \Rightarrow \Phi(F)}$$

But what might be a suitable choice for 'Ψ'? The existential quantifier would be safe enough in the role, but would once again leave us without the resources to instantiate on legitimate but unsatisfied predicates. What *is* the quantifier, actually, that says of a predicate F that its sense and the world have so cooperated as to make it true that there is such a thing as being F?

The issue—to give a proper account of the distinction for predicates, unexampled in the case of singular terms, between (something akin to) lacking reference and being unsatisfied—is of some independent interest.[9] Once properly clarified, there would be no objection to introducing a primitive second-level predicate to mark it, and framing the UE rule in terms of that. But since it is, of course,

[9] One suggestion (put forward by Bob Hale in correspondence) is that an assurance against the counter-

an informal philosophical distinction, we would be bound to acknowledge a consequential compromise in the ideal of a fully rigorous syntactic capture of the rules (though admittedly a kind of compromise we already make in the formulation of free first-order logic). My present point again, however, is merely that it should not necessarily be regarded as compromising the generality of a neutralist account of Universal Elimination if the details for the second-order case turn out to be somewhat different from those appropriate in the first-order case. What is crucial is the common pattern of inference and that the additional constraints on that pattern, respectively appropriate in the first- and second-order cases, should be intelligible as addressed to common requirements: viz. genuine generality, and the need—where the languages in question may throw up examples of such—to address the risk of (something akin to) reference failure.

3 *Extreme neutralism*

Let me stress again the sense in which what is aimed at is a neutralist account of quantification. The central thesis is not that higher-order quantification ought to be construed in such a way that it implicates no special ontological commitment. Rather it is that quantification is not *per se* ontologically committing: Quine's idea, that the ontological commitments of a theory are made fully manifest by the range of kinds of quantification that a minimally adequate formulation of the theory requires—the idea rather unhappily encapsulated in the slogan 'to be is to be the value of a variable'—is simply a mistake, encouraged by a preoccupation with first-order quantification in a language all of whose names refer. The Neutrality Principle has it that statements resulting from quantification into places occupied by expressions of a certain determinate syntactic type need not and should not be conceived as introducing a type of ontological commitment not already involved in the truth of statements configuring expressions of that type. In particular, if there is a good account of the nature of predication which frees it of semantic association with entities such as universals, properties or concepts, then quantification into predicate position does not introduce any such commitment.

Again: quantifiers are merely devices of generalization—devices for generating statements whose truth-conditions will correspond in ways specific to the kind of quantifier involved to those of the statements quantified into—the instances. Applied to second-order quantifiers in particular, the neutralist thesis is thus that they do not differ from those of first-order logic by the implication of a type of ontological commitment not already present in the former. If second-order quantifiers demand a special kind of—perhaps problematic—entity, that will be because that same demand is already present merely in the idea of atomic predication in which, of course, in its characteristic recourse to predicate-schemata,

part of reference-failure for predicates is provided by satisfaction of the generalized Law of Excluded Middle, $(\forall x)(Fx \vee \neg Fx)$. But one would need some principled way to head off the response of an awkward customer that if John has no favourite colour, then it is true of my shirt that it is not John's favourite colour.

first-order logic is already up to its eyeballs. There is thus, in the neutralist view, no interesting difference in the ontological commitments of first- and second-order quantification, and no interesting difference between the ontological commitments of either and those of the quantifier-free atomic statements to which they are applied. If there is a residual sense in which only first-order logic is properly styled as logic, it is not at any rate an additional implicit ontology that disqualifies second-order logic from that honorific title.

We should note that the spirit of this proposal is consistent with a particularly extreme form of incarnation. From a strict neutralist point of view, there can be no *automatic* objection to existential quantification into the place occupied by 'the largest number' in

(a) There is no such thing as the largest number.

Plainly 'the largest number' is not being used referentially in that sentence. So its existential generalization, which merely involves generalization of whatever its semantic role there is, will not—absurdly—imply the existence of something which there is no such thing as. What it will imply is merely that some statement is true for whose truth it suffices that some—it doesn't matter which—statement of the form

There is no such thing as #

be true, where '#' holds a place for an expression whose semantic role is that of 'the largest number' in (a). Regular—existence implying—first-order existential quantification will thus be a special case of neutral first-order existential quantification, falling out of instances where the term quantified into *is* functioning referentially (i.e. functioning in such a way that its failure to refer will divest the containing context of truth).

I think we should grant that this extreme neutralism is a prima facie defensible option. Of course it may run into trouble, but there is nothing in neutralism *per se* that closes it off. There are good reasons, nevertheless, to fashion an account of the quantifiers which allows the connections, in the first-order case, with the idea of a range of entities, and with ontological commitment, which Quine took to be central. In that case, we will prefer—for languages involving possibly empty terms—to employ something like the approach illustrated by the free-logical rule proposed for UE above.

4 *A neutralist heuristic*

So to recap: our proposal is that the meanings of the quantifiers of all orders, neutrally conceived, are to be viewed as grounded in the inferential practices licensed by their introduction and elimination rules, that they are as logical as any other concepts of comparable generality which are associated in this way with a distinctive role in the inferential architecture of subject-predicate thought as such, and that there is no better reason to view them as associated with distinctive ontological commitments than there is so to regard the conditional or negation. Still, old habits die hard and the temptation may remain to press for an informal elucidation of what, in using operators so characterized, we are really *saying*—what kind of claim is being made by someone who offers a higher-order generalization if he is not—or need not—be 'quantifying over' properties or the like? The felt need remains for a *heuristic*—a way of thinking about the gist of higher-order generalization that frees it of the Quinean associations and makes some kind of sense of the practices codified in the rules. We run a risk, of course, in offering such a heuristic, of confusing it with a theory of the actual meaning of the statements at issue—a risk, by the way, which is also run by formal model theory. Such a heuristic, or model theory, may very well incorporate ideological and ontological assumptions which are, properly speaking, no part of the content of the targeted class of statements. Still the result may be helpful. No one can deny the utility and fruitfulness of possible-worlds semantics, for example, as a model for modal discourse, but relatively few believe that possible-worlds semantics uncovers its actual content. I acknowledge, naturally, that there is an issue of wherein the utility and fruitfulness consists if actual meanings are not recovered. I'll make a suggestion about that shortly.

What can be offered by way of an informal heuristic for quantification when neutrally viewed? Here is one suggestion. Begin with *substitutional* quantification as normally conceived. A substitutional quantifier is assigned a range of significance with respect to a given determinate language: we first individuate a certain grammatical type of expression, and then specify that, for instance, the result of existential generalization into a place in a particular statement occupied by an expression of that type is to be true just in case some sentence of the language differing from that statement only at most by the occurrence in that place of a different expression of the same grammatical type is a true sentence. The corresponding explanations for other kinds of quantifiers proceed in the obvious way. Two salient points are these: first, that in characterizing the truth-conditions of statements formed by substitutional quantification, we resort—in the meta-language—to *non-substitutional* quantification over *sentences* in the object language; and second, and very familiarly, that the scope of the generality that substitutional quantifiers enable us to achieve is bounded by the actual vocabulary—in particular, by the range of expressions of the particular grammatical kind concerned—of the object language in question. Hence the familiar complaint that the generality provided by substitutional quantification is not the true generality that quantification should involve.

The neutralist heuristic I have in mind takes off from these two points. First, in explicating the truth-conditions of quantified sentences, we now resort—in the meta-language—to quantification not over sentences but over the things that sentences express: over *thoughts* (propositions). Second, we waive the restriction to particular (limited) languages. We conceive of thoughts as structured entities with a structure corresponding to that of the deep-semantics of a properly canonical expression for them, and quantification is then seen as a device of generalization into the positions configured in structured entities of this kind.[10] Thus, for instance, the existential generalization of any significant syntactic constituent in the expression of a thought takes us to a thought which is true just in case some thought is true differing at most only in the syntactic constituent that would figure in that place in a canonical expression of it. Correspondingly for universal quantification.

This is a notion of quantification whose generality is unbounded by the expressive resources of any particular actual language—it provides for generalized commitments in a way which embraces arbitrary *extensions* of such a language—and which manifestly carries only the ontological commitments already carried by the use of the syntactic type of expression quantified into. However, it is true that the heuristic also involves quantification over *thoughts*. That may provoke the misgiving that it parts company with the neutralist intent at the crucial point and so misses its mark. For was not the idea to explicate an ontologically non-committal conception of higher-order quantification, free of any association with a range of entities?

Not exactly. Rather, the idea was, in the first instance, to explicate a conception of quantification into predicate—or indeed into any syntactic kind of—position which would avoid associating occurrences of expressions of that syntactic kind with a *distinctive kind* of entity, presumed to provide their semantic values. According to the heuristic, all quantification, of whatever order, into whatever syntactic kind of place in a statement, is quantification 'over'—generalization with respect to—thoughts. So the heuristic saves a key element of the *Begriffsschrift* conception—its indifference to the distinctions of order that, after Quine, came to seem so crucial.

But there is a more fundamental point. The examples of regular set-theoretic semantics for first-order languages and of possible-worlds semantics for modal languages remind us of the point I mentioned a moment ago, that model theory, whether in formal, rigorous form or that of informal heuristics, need not—and often had better not—be thought of as conveying the actual content—the *actual*

[10] In so conceiving of thoughts as structured I of course follow the lead of Evans in the following memorable passages:

> It seems to me that there must be a sense in which thoughts are structured. The thought that John is happy has something in common with the thought that Harry is happy, and the thought that John is happy has something in common with the thought that John is sad. This might seem to lead immediately to the idea of a language of thought ... However, I certainly do not wish to be committed to the idea that having thoughts involves the subject's using, manipulating or apprehending *symbols*. ... I should prefer to explain the sense in which thoughts are structured, not in terms of their being composed of several distinct elements. but in terms of their being a complex of the exercise of several distinct conceptual *abilities*. ... While sentences need not be structured, thoughts are essentially structured. Any meaning expressed by a structured sentence could be expressed by an unstructured sentence. ... But it is simply not a possibility for the thought that *a* is *F* to be unstructured—that is, [for the thinking of it] not to be the exercise of two distinct abilities. (Evans 1982: 100–102)

conceptual resources deployed in—the statements of the explicated language. In the present instance, that content remains fixed operationally by the introduction- and elimination-rules for the quantifiers. What the heuristic does is provide an informal theory which, as it were, maps, or reflects, the truth-conditions of the statements whose contents are so fixed without pretending to recover the understanding of them possessed by competent speakers. Although it would take us too far afield to explore the idea thoroughly here, I suggest that the right way to think of this mapping relationship is that it consists in an identity of truth-conditions *modulo* the assumption of the ontology of the heuristic or model theory. The very simplest example of such a relationship is provided by the relation between the statement that Socrates is a man and the statement that Socrates is a member of the set of men—a relation of necessary equivalence in truth-conditions *modulo* a modest ontology of sets (sameness of truth value in all worlds, if you will, in which the ontology of the model theory is realized). The same view may be held concerning the statements that the number of planets is the same as the number of clementines in the fruit-bowl and that there are exactly as many planets as clementines in the fruit-bowl—a necessary equivalence in truth-conditions *modulo* the ontology of the finite cardinals.[11] And it is, so I would argue, the same with unreconstructed modal claims and their construals in possible-worlds semantics. In each case, a theoretical explication of content is achieved via an equivalence in truth-conditions under the hypothesis of the ontology of the explicating theory. Where T is a statement of the theory concerned, S a statement whose content is to be explicated, and E its explicans, the explicating relationship is of the schematic form

$$T \Rightarrow (S \Leftrightarrow E),$$

and explication is achieved, when it is, via the (conceptual) necessity of appropriate instances of this schema.

The distinction between explication of this kind and some yet closer relation of content that we might want to bring under the old fashioned term 'analysis' is something we need generally if we are to understand the proper role—both the prospects and the limitations—of interpretative semantic theory in philosophy. It is a distinction that is, ironically, enforced by the precedent of set-theoretic semantics for first-order theories yet completely missed by the normal, Quinean way of receiving higher-order logics' standard set-theoretic semantics. My suggestion would be that we avail ourselves of it, in the form I have adumbrated, to understand the relationship between the sketched heuristic and (higher-order) quantification: a quantified statement and the kind of paraphrase we may give of it in terms of the heuristic are equivalences *modulo* the ontology of a suitable theory of thoughts as structured entities—a theory which, speaking for myself, I think there is ample reason to endorse and to develop but of which it is open to us to take an agnostic, or even fictionalist view, while still deriving illumination from the content mappings it lets us construct.

[11] Of course the abstractionist view of this particular equivalence is that it holds unconditionally.

5 Comprehension

In standard formulations of higher-order logic, comprehension axiom are used to tell us which predicates—open sentences—are associated with an entity belonging to the domain of quantification in question and are thus safe for quantification. They are therefore to be compared with postulates in a free first-order logic telling us which of the terms in a language in question have reference to an object. It goes with neutralism, as I've been outlining it, that there need be—more accurately: that for extreme neutralism, at least, there is—*no role* for comprehension axioms. If quantification is viewed as generalization of semantic function, whatever it is, where discharging the semantic function in question just means contributing appropriately to a range of intelligible thoughts, there can be no hostage, comparable to reference failure, which must be redeemed before quantification is safe. If an expression belongs to a determinate syntactic kind, and plays the semantic role appropriate to an expression of that kind in the canonical expression of a given intelligible thought, then quantification will be safe without additional assumption.

Still, one will need circumspection about its *significance*, which will be constrained by the actual semantic role of the expression quantified into in the statement concerned. Existential quantification, for instance, as we have noted, may not *per se* carry the kind of existential commitment with which it is normally associated. But one may still wish to make sure that it does, that only *referential* singular terms are eligible for existential generalization. And, as we have noted, there may be similarly desirable restrictions to be placed on those expressions which are to be eligible for higher-order generalization. The informal example of '...is John's favourite colour' is a reminder that, once we go past the atomic case and allow ourselves the luxury of forming new predicates by quantification into or other forms of operation (in this case, definite description) on predicates, we may find ourselves forming significant expressions which nevertheless stand in some kind of parallel to referenceless singular terms.

There is much more to say, but we can, I think, now table a group of natural proposals about the specific form that introduction and elimination rules for second-order universal and existential quantification may assume consonantly with an inferentialist conception of the meaning of those operations and the neutralist heuristic. The universal rules are, unremarkably,

UE $\dfrac{\Gamma \Rightarrow (\forall X)\Phi X}{\Gamma \Rightarrow \Phi(F)}$, where '$F$' is good

UI $\dfrac{\Gamma \Rightarrow \Phi(F)}{\Gamma \Rightarrow (\forall X)\Phi X}$, for arbitrary atomic '$F$'

while those for the existential quantifier are

EI $\Gamma \Rightarrow \Phi(F)$, where '$F$' is good
$\Gamma \Rightarrow (\exists X)\Phi X$

EE $\Gamma \Rightarrow (\exists X)\Phi X$; $\Delta, \Phi(F) \Rightarrow P$, for arbitrary atomic '$F$'
$\Gamma, \Delta \Rightarrow P$

Departures from extreme neutralism will be signalled by activating the 'where "F" is good' clause, while the effect of one's favourite comprehension restrictions may be achieved by specific interpretations of what goodness consists in. The effect of full, impredicative second-order comprehension may be accomplished by counting all meaningful open sentences as good, save possibly those afflicted by (the analogue of) reference-failure, and by finding no obstacle to meaningfulness in the free formation of open sentences involving the second-order quantifiers themselves. More on this below.

We should note at this point, however, that the *generosity* of standard Comprehension axioms in classical higher-order logic is hard to underwrite by means of the proposed heuristic. A standard formulation of classical Comprehension will provide that, for each formula Φ of the language in n argument places, there is a relation in n places, X, such that for any n-tuple $\langle x_1, \ldots, x_n \rangle$, X holds of $\langle x_1, \ldots, x_n \rangle$ just in case $\Phi\langle x_1, \ldots, x_n \rangle$. At first blush, this seems perfectly consonant with neutralism: existential generalization into predicate position is legitimated just when we have an intelligible formula of the language (or some extension of it) to provide for a witness. But the appearance is deceptive. Comprehension as standardly formulated has, for example, this instance:

$$(\forall x)(\exists X)(\forall y)(Xy \leftrightarrow y = x)$$

—intuitively, that for any thing, there is a property (that is, a way things can be) which a thing exemplifies just in case it is identical with that thing. So we get a range of properties specified simultaneously as a bunch, in a way that is parametric in the variable 'x'. This is consonant with a connection between second-order quantification and the intelligibility of the thoughts that constitute its instances only if each object in the first-order domain in question—each object in the range of 'x'—is itself a possible object of intelligible singular thought.

To see the point sharply, consider how we might set about establishing the above instance of comprehension using the higher-order rules outlined above and the standard first-order rules. Presumably we have as a theorem at first-order:

$$(\forall x)(\forall y)(y = x \leftrightarrow y = x).$$

So instantiating on 'x', we obtain

$$(\forall y)(y = a \leftrightarrow y = a).$$

Then, assuming '$\ldots = a$' is good, the proposed **EI** rule allows the move to

$$(\exists X)(\forall y)(Xy \leftrightarrow y = a).$$

And finally, since each of these theses is justified purely by the logic, and 'Γ' is empty throughout, we can generalize on 'a' to arrive at the target claim:

$$(\forall x)(\exists X)(\forall y)(Xy \leftrightarrow y = x).$$

Manifestly, however, the way we have arrived there depends entirely on the good standing of the open sentence '... $= a$', which in turn is naturally taken to presuppose the existence of a suitable singular mode of presentation of the object concerned. The 'generous' instance of Comprehension is established only relative to that presupposition.

With uncountably infinite domains, however, the presupposition is doubtful. Only an infinite notation could provide the means canonically to distinguish each classical real number from every other in the way that the standard decimal notation provides the means canonically to distinguish among the naturals. So, on plausible assumptions, no finite mind can think individuative thoughts of every real. Yet the instance of Comprehension cited implies that to each of them corresponds a distinctive property. So standard Comprehension strains the tie with intelligible predication, crucial to the neutralist heuristic as so far understood—at least, it does so if classically uncountable populations of objects are admitted. Quantification over uncountable totalities admits of no evidently competitive interpretation in terms of distribution of truth-values through a range of intelligible singular thoughts.

That there would likely be difficulties when it comes to reconciling a broad conception of quantification as distributive of truth-values among intelligible thoughts with the classical model-theoretic (set-theoretic) conception of higher-order quantification was of course obvious from the start, since the classical conception associates each nth order of quantification with a domain of 'properties' (sets) of the nth cardinality in the series of Beth numbers and nothing resembling human thought, nor any intelligible idealization of it, is going to be able to encompass so many 'ways things might be'. But the point developed above concerns uncountable domains of *objects*, not higher-order domains, and calls into question whether the kind of conception of quantification which I am canvassing as possibly suitable for the abstractionist programme in the philosophy of mathematics is actually fit for purpose—assuming that the project is to embrace the pure mathematics of certain uncountable domains, including *par excellence* the classical real numbers. To stress: the difficulty is with the heuristic, not (yet) with the inferentialist conception of quantification *per se*. But there is clearly matter here for further study and invention.

6 *Incompleteness*

There is a salient objection to the neutralist approach—indeed to any inferentialist account of higher-order quantification—which we can defer no longer. It springs from Gödel's results about the incompleteness of arithmetic. Here is an intuitive statement of it.[12]

As is familiar, second-order arithmetic is, like first-order arithmetic, shown by Gödel's findings to be incomplete. But there are crucial differences. The completeness of first-order logic ensures that every sentence that holds in every model of first-order arithmetic can be derived from the first-order Peano Axioms by its deductive resources. So the incompleteness of first-order arithmetic means that some truths of arithmetic do not hold in every model of the first-order Peano Axioms. However the second-order Peano axioms are categorical: they have exactly one (standard) model and any truth of arithmetic holds in it. Hence the incompleteness of second-order arithmetic entails that second-order logic is incomplete: not every sentence that holds in every (i.e. up to isomorphism, the unique) model of the second-order Peano Axioms can be derived from them in second-order logic. So does this point not somehow enforce a model-theoretic perspective on the idea of second-order logical validity? For surely we are now forced to allow that there are valid second-order logical sentences which cannot be established by second-order deduction. Yet what sense can be made of this if we are inferentialist about the meanings of the quantifiers? How can we regard the meanings of the quantifiers as fully fixed by the inference rules and at the same time allow that there are valid sentences expressible using just second-order logical vocabulary which those rules fail to provide the means to establish?

Well, are we forced to say this? Let's go carefully. Let 2PA be a suitable axiom set for second-order arithmetic, and let G be the Gödel sentence for the resulting system. Let 2G* be the conditional, 2PA → G. Clearly this cannot be proved in 2PA. And the usual informal reasons for regarding G as a truth of arithmetic weigh in favour of regarding this conditional as a truth of arithmetic. But they are not reasons for regarding it as a truth of logic—it is shot through with non-logical vocabulary. However consider its universal closure U2G*. This is expressed purely second-order logically but it too cannot be provable in 2OL, for if it were, we would be able to prove 2G* in 2PA. Is there any reason to regard U2G* as a logical truth/validity? If there is, that will be a prima facie reason to regard second-order validity as underdetermined by the second-order inference rules, including those for the quantifiers—a body-blow, seemingly, against inferentialism.

There is indeed such a reason. *Third-order* logic allows us to define a truth-predicate for second-order logic and thereby to mimic rigorously in a formal third-order deduction the informal reasoning that justifies the conclusion that G holds good of any population of objects that satisfy 2PA. So 2G* can be proved using just 3OL resources, without special assumption about the arithmetical primitives. There is therefore also a legitimate generalization to U2G*. So now we have a sentence expressible by means purely of second-order logical vocabulary, and establishable as a theorem of third-order logic but underivable in second-order

[12]The discussion to follow is greatly indebted to discussion with Marcus Rossberg.

logic.[13] If third-order logic is logic, this sentence is a logical truth that configures no logical concepts higher than second-order yet cannot be established by the inference rules of 2OL.

The situation generalizes. From a non-model-theoretic perspective, the incompleteness entailed by Gödel's results for each *n*th order of higher-order logics consists in the deductive non-conservativeness of its hierarchical superiors over it—each *n*th-order logic provides means for the deduction of theorems expressible using just the conceptual resources of its predecessor but which cannot be derived using just the deductive resources of its predecessor. But then at no *n*th order is it open to us to regard the meanings of the quantifiers at that order as fully characterized by the *n*th-order rules—otherwise, what could possibly explain the existence of truths configuring just those meanings but underivable by means of the rules in question?

There are various possible responses. One might attempt a case that 3OL is *not* logic, so that some of its consequences, even those expressed in purely second-order logical vocabulary, need not on that account be rated as logical truths.[14] But the prospects for this kind of line look pretty forlorn in the present context—after all, the neutralist conception of quantification embraces *any* kind of open sentence, including in particular any resulting from the omission from complete sentences of first-order predicates. The resulting class of expressions ought to be open to generalization in just the same sense as first-order predicates. How could the one kind of operation be 'logical' and the other not?

However, I think it is the very generality of the neutralist conception of quantification that points to the correct inferentialist response to the problem. That response is, in essentials, as follows. Epistemologically, it is a mistake to think of higher-order quantifiers as coming in conceptually independent layers, with the second-order quantifiers fixed by the second-order rules, the third-order quantifiers fixed by the third-order rules, and so on. Rather it is the *entire series* of pairs of higher-and higher-order quantifier rules which collectively fix the meaning of quantification at each order: there are *single* concepts of higher-order universal and existential generalization, embracing all the orders, of which it is possible only to give a schematic statement—as indeed we effectively did in the rules proposed earlier, in which, although I then announced them as 'second-order', and accordingly represented the occurrences of 'F' within them as schematic for predicates of objects, no feature was incorporated to bar the construal of 'F' as schematic for predicates of any particular higher-order, including—if one is pleased to go so far—transfinite ones. Higher-order quantification is a uniform operation, open to a single schematic inferentialist characterization, and there is no barrier to regarding each and every truth of higher-order logic as grounded in the operation of rules that are sanctioned by that characterization.

[13] For details, see section 3.7 and 4.1 of Leivant (1994).

[14] By way of a parallel, the at least countable infinity of the universe may be expressed in purely logical vocabulary. Still, it may coherently be regarded as a necessary *mathematical* truth rather than a logical one.

7 *Impredicativity*

Classical second-order logic allows full *impredicative* comprehension. Impredicative comprehension is this process. In an open sentence, '$\ldots F \ldots$', configuring one or more occurrences of a (simple or complex) predicate, we first form a new predicate by binding one or more occurrences of 'F' with a quantifier—say '$(\forall X) \ldots X \ldots$'—and then treat this new predicate as falling within the scope of its own quantifier. The manoeuvre is essential to a number of fundamental theorems in classical foundations, including the second-order logical version of Cantor's theorem and various key lemmas in the abstractionist recovery of arithmetic and analysis.

Can neutralism underwrite impredicative quantification? The standard objection to impredicative quantification of all orders is epistemological—that it introduces a risk of vicious circularity, or ungroundedness, into the meaning of expressions which involve it. One reason for thinking this which we may discount for present purposes is endorsed in various places by Michael Dummett.[15] Dummett's objection is that determinacy in the truth-conditions of statements formed by means of quantification depends upon a prior definite specification of the range of the quantifiers involved, and that such a definite specification may be impossible to achieve if the quantifiers are allowed to be impredicative, since their range may then include items which can only be specified by recourse to quantification over the very domain in question. I think *this* form of worry involves a mistake about the preconditions of intelligible quantification. It simply isn't true that to understand a quantified statement presupposes an understanding of a domain specified independently—as is easily seen by the reflection that such an independent specification could hardly avoid quantification in its turn.[16] But even waiving that, Dummett's objection can hardly be germane in a context in which we are precisely trying to view higher-order quantification as *range-free*. Range-free quantification involves no domain.

So is there a problem at all? Well, we can take it that if the quantifiers allow of satisfactory inferential role explanations at all, then higher-order quantified sentences will be intelligible and have determinate truth-conditions whenever their instances do. So any genuine problem to do with impredicativity here must have to do with the intelligibility of the enlarged class of instances that it generates—the new *open sentences* that impredicative quantification allows us to formulate. The problem, if there is one, will have to concern whether or not such sentences have determinate satisfaction-conditions.

There is a lot to think about here, but it is clear—at least insofar as the project is informed by the proposed heuristic—that there is no good objection to one basic kind of case. Consider this example.

[15]See especially various passages in ch. 18, 'Abstract Objects', of his (1991).

[16]A point emphasized by (Hale 1994). The issues are further discussed in my (1998). Both papers are reprinted in Hale and Wright (2001).

Federer is tactically adroit
Federer has pace and stamina
Federer has a complete repertoire of shots
Federer is gracious in victory

Federer is everything a tennis champion should be

The last is naturally understood as an impredicative second-order quantification—after all, being everything a champion should be is certainly one thing, among the rest, that a champion should be! But notice that this introduces no indeterminacy into the satisfaction-conditions of '... is everything a tennis champion should be' since it remains necessary and sufficient for the application of that predicate that a player have a range of *predicatively specifiable* characteristics, like those illustrated by the first four items in the list. That is, it is necessary and sufficient for Federer to be everything a tennis champion should be—including that very characteristic—that he have a range of *other* characteristics, not including that characteristic, whose satisfaction-conditions can be grasped independently.

The example is one of a predicate formed by means of impredicative quantification which is fully intelligible because its satisfaction-conditions can be explicated without recourse to impredicative quantification. Russell in effect tried to assimilate all cases of impredicative quantification to this model by his famous Axiom of Reducibility, which postulated that every (propositional) function of whatever kind is equivalent to a predicative function—one in whose specification no higher-order quantification need be involved. But of course he was unable to provide a general compelling motive for believing the Axiom. The point remains that impredicative quantification need not *per se* pose any special obstacle for neutralism. The crucial question for second-order impredicative quantification in particular is just whether allowing predicates formed by impredicative quantification results in predicates of whose satisfaction-conditions there is no clear (non-circular) account. It is clear that there is no reason in general to suppose that has to be so. When it is not so, such new predicates will be open to intelligible neutral quantification in turn.

But how general is that run of cases? Do all the important examples of impredicative quantification allowed in classical higher-order logic fall on the right side of that distinction? What about the proof (strictly in third-order logic, I suppose) of the second-order logical version of Cantor's Theorem? (And how, on a range-free conception of quantification, is that proof to be interpreted in any case?) Do all the types of impredicative quantification demanded by the abstractionist programme fall on the right side of the distinction? Frege's definition of the Ancestral of a relation is impredicative. What are the implications of that for its role in the abstractionist programme?

We need answers to these questions, but clear answers will take some digging out. Of course, issues about impredicativity have long been regarded as pivotal in the debates about the foundations of mathematics for more than a century.

But the discussion of them has traditionally been taken in the context of—and clouded by—the opposition between platonist and constructivist philosophies of mathematics. Not the least of the merits of the neutralist view, it seems to me, is that it lets us relocate these issues where they belong, in the proper account of quantification and the epistemology of logic.

I have offered the beginnings of a case that there is a legitimate conception of higher-order quantification which views it, like first-order quantification, as an operation grounded purely in the structure of subject-predicate thought and consequently as logical in whatever sense first-order quantification is logical. I confidently assert that *some* version of second-order logic, so conceived, is pure logic if anything is pure logic. But the question remains how much of *classical* second-order logic can be recovered under the aegis of this conception, when properly worked out, and whether the resulting system(s) can be meet for the needs of technical philosophical programmes, such as Abstractionism or Conceptual Realism based on categoricity results.

8 *Appendix: Abstractionist Mathematical Theories*

On a thumbnail, the technical part of the Abstractionist project is to develop branches of established mathematics in higher-order—in practice, second-order—logic, with sole additional axioms consisting of *abstraction principles* (often simply termed *abstractions*). These are axioms of the form

$$\forall a \forall b(\Sigma(a) = \Sigma(b) \leftrightarrow E(a, b)),$$

where a and b are variables of first- or higher-order (typically, second-order), 'Σ' is a singular-term forming operator denoting a function from items in the range of 'a' and 'b' to objects, and E is an equivalence relation over items of the given type.[17]

Arithmetic

Hume's Principle is the abstraction:

$$\forall F \forall G(Nx : Fx = Nx : Gx) \leftrightarrow (F \text{ 1-1 } G),$$

where 'F 1-1 G' is an abbreviation of the second-order statement that there is a one-to-one relation mapping the Fs onto the Gs. Hume's Principle thus states that the number of Fs is identical to the number of Gs if and only if F stands in such a relation to G. Adjoined to a suitable impredicative higher-order logic, it is technically sufficient for classical number-theory.

[17] Of course, the conception of S as denoting an operation on *things* of a certain order—in general, the orthodox conception of a higher-order function—and the idea of E as an equivalence relation on such things need to be rethought if one is going in for a range-free conception of higher-order quantification along the lines of the present discussion.

Real Analysis

Real analysis can be obtained by starting with Hume's Principle plus full, impredicative second-order logic. We use the *Pairs* abstraction:

$$(\forall x)(\forall y)(\forall z)(\forall w)(\langle x,y\rangle = \langle z,w\rangle \leftrightarrow x = z \wedge y = w)$$

to get the ordered pairs of the finite cardinals so provided. An abstraction over the *Differences* between such pairs:

$$\mathrm{Diff}(\langle x,y\rangle) = \mathrm{Diff}(\langle z,w\rangle) \leftrightarrow x + w = y + z,$$

provides objects with which we can identify the *integers*. Define addition and multiplication on these integers. Next, where m, n, p and q are any integers, we form *Quotients* of pairs of integers in accordance with this abstraction:

$$Q\langle m,n\rangle = Q\langle p,q\rangle \leftrightarrow (n = 0 \wedge q = 0) \vee (n \neq 0 \wedge q \neq 0 \wedge m \times q = n \times p),$$

identifying a *rational* with any quotient $Q\langle m,n\rangle$ whose second term n is non-zero. Define addition and multiplication and thence the natural linear order on the rationals so generated. Move on to the Dedekind-inspired *Cut Abstraction*:

$$(\forall F)(\forall G)(\mathrm{Cut}(F) = \mathrm{Cut}(G) \leftrightarrow (\forall r)(F \leq r \leftrightarrow G \leq r)$$

where 'r' ranges over rationals, and the relation '$\leq$' holds between a property F of rationals and a specific rational number r just in case any instance of F is less than or equal to r under the constructed linear order on the rationals. Identify the *real numbers* with the cuts of those properties F, G which are both bounded above and instantiated in the rationals.[18] It can be shown that the reals so arrived at constitute a complete, ordered field, exactly as is classically conceived as distinctive of them.

Set Theory

No comparable abstractionist successes can be documented in this area so far. The best researched proposal for a set theory founded on abstraction takes the usual second-order logical setting with a single abstraction principle, George Boolos's so-called 'New V':

$$\forall F \forall G(\{x : Fx\} = \{x : Gx\}) \leftrightarrow ((\mathrm{Bad}(F) \wedge \mathrm{Bad}(G)) \vee (\forall x)(Fx \leftrightarrow Gx))$$

—informally, the abstracts of F and G are the same just in case either F and G are both bad, or are co-extensive. Here, 'badness' is identified with having the size of the universe—in effect, with sustaining a bijection with the universal concept—and the effect of the proposal, in keeping with the tradition of limitation of size, is to divide the abstracts it yields into well-behaved ones, corresponding to 'good'

[18]This bracing progression is developed by Shapiro (2000). An alternative, closer in spirit to Frege's insistence that a satisfactory foundation for a mathematical theory must somehow integrate its potential applications into the basic explanations, is elegantly developed by Hale (2000*b*).

concepts, for which identity is given by co-extensionality, as for 'naive' sets, and a single bad object corresponding to all the universe-sized concepts whether co-extensive or not. We can think of the *sets* as the former group.

As Boolos (1987; 1989) shows, a surprising number of the laws of classical set-theory can be verified of the sets (though not necessarily holding in the full domain) delivered by New V. Extensionality, Null set, Pairs, Regularity, Choice (global), Replacement, Separation, and Union are all forthcoming. But neither Infinity—that is, the existence of an infinite *set* (as oppose to the infinity of the domain)—nor Power Set can be obtained.

How these crucial shortcomings might be remedied—if indeed any satisfying remedy is possible at all—is a matter of current research. It will be evident that in order to obtain an infinite set via New V we need first to obtain a concept larger than that by reference to which the sought-for set is to be defined; at a minimum, then, we first need an uncountable domain in order to generate a countably infinite set. As for Power Set, the requirement becomes that, for any infinite set so far obtained, a concept can be found large enough to outstrip the cardinality of its power set: that is, for any F, there exists a G such that the sub-concepts of F, as it were, can be injected into G but not vice versa. A natural thought is to try to augment the comprehension afforded by New V with the objects generated by an abstraction affording an uncountable domain, for example the Cut abstraction principle given above. A principle of this form may be given for any densely ordered infinite domain, and will always generate more cuts than elements in such a domain—indeed exactly 2^n many cuts if n is the cardinality of the original domain. So if we could show that any infinite concept allowed of a dense order, and could justify limitless invocation of appropriate Cut principles, then Cut abstraction would provide resources to generate limitlessly large concepts, and New V would generate limitlessly large well-behaved sets in tandem.

But the provisos are problematic. There are issues about abstractionism's right to the limitless invocation of Cut principles.[19] And the crucial lemma, that any infinite domain of objects allows of a dense order, is standardly proved[20] in ways that presuppose the existence of cardinals—hence sets—larger than the domain in question: the very point at issue. Finally, New V itself is anyway in violation of certain of the conservativeness constraints to which abstractions arguably ought to be subject, by virtue of its very entailment of the particularly strong form of Choice that it delivers—the global well-ordering of the universe (including, of course, any non-sets that it contains).[21] Although it would certainly be a mistake to dismiss the prospect of any significant contribution by abstractionism to the epistemology of set theory,[22] my own suspicion is that the inability of the

[19] See the exchange between Cook (2002) and Hale (2000*a*).

[20] For example in Cook (2002).

[21] The downside of this result is emphasized in Shapiro and Weir (1999).

[22] A reader who wants to think further about the prospects for an abstractionist recovery of a strong

approach to justify more than a tiny fraction of the conventionally accepted ontology of sets may prove to be an enduring feature. Abstraction principles articulate identity-conditions for the abstracts they govern, and so explain what kind of things they are. It is down to the characteristics of the abstractive domain—the field of the equivalence relation on the right hand side—how much comprehension is packed into this explanation. If it turns out that any epistemologically and technically well-founded abstractionist set theory falls way short of the ontological plenitude we have become accustomed to require, we should conclude that nothing in the nature of sets, as determined by their fundamental grounds of identity and distinctness, nor any uncontroversial features of other domains on which sets may be formed, underwrites a belief in the reality of that rich plenitude. The question should then be: what, do we suppose, does?[23]

set-theory (perhaps, at a minimum, something equivalent to ZFC) should turn—in addition to the sources referenced in other notes in the present essay—to Shapiro (2003), Fine (2002), and Burgess (2005).)

[23]Thanks to various sometime members of the Arché *Logical and Metaphysical Foundations of Classical Mathematics* project—particularly Philip Ebert, Robbie Williams, Nikolaj Pedersen, Marcus Rossberg, Stewart Shapiro, William Stirton, Bob Hale, Roy Cook, Fraser MacBride, Agustín Rayo, and Elia Zardini—for helpful discussion of ancestors of this material presented at Arché seminars in 2003 and 2004, and for critical reactions since.

Bibliography

Baillargeon, R. (2000), 'Reply to Bogartz, Shinskey, and Schilling; Schilling; and Cashon and Cohen', *Infancy* **1**, 447–62

Baker, Alan (2001), 'Mathematics, indispensability and scientific progress', *Erkenntnis* **55**, 85–116

— (2003), 'The indispensability argument and multiple foundations for mathematics', *Philosophical Quarterly* **53**, 49–67

— (2005), 'Are there genuine mathematical explanations of physical phenomena?', *Mind* **114**, 223–37

Balaguer, Mark (1998), *Platonism and Anti-Platonism in Mathematics*, Oxford University Press

Beebee, Helen (2001), 'Transfer of warrant, begging the question, and semantic externalism', *Philosophical Quarterly* **51**, 356–74

Bell, David (1988), 'Phenomenology and egocentric thought', *PASSV* **62**, 45–60

Benacerraf, Paul (1973), 'Mathematical truth', *Journal of Philosophy* **70**, 661–80

Blackburn, Simon (1994), *The Oxford Dictionary of Philosophy*, Oxford University Press

Boolos, George (1984), 'To be is to be the value of a variable (or to be some values of some variables)', *J. Phil.* **81**, 430–49

— (1987), 'Saving Frege from contradiction', *Proc. Arist. Soc.* **87**, 137–51

— (1989), 'Iteration again', *Phil. Topics* **42**, 5–21

Boroditsky, L. (2001), 'Does language shape thought? Mandarin and English speakers' conceptions of time', *Cognitive Psychology* **43**, 1–22

Brandstein, M. (1982), 'New lower bound for a factor of an odd perfect number', *Abstracts Amer. Math. Soc.* **3**, 257

Brent, R. P., G. L. Cohen, and H. J. J. te Riele (1991), 'Improved techniques for lower bounds for odd perfect numbers', *Mathematics of Computation* **57**, 857–68

Brown, Jessica (2003), 'The reductio argument and transmission of warrant', in S. Nuccetelli (ed.), *New Essays on Semantic Externalism and Self-Knowledge*, MIT Press, Cambridge, MA, pp. 117–30

Bueno, Otávio, and Mark Colyvan (n.d.), An inferential conception of the application of mathematics. Forthcoming

Burge, Tyler (1993), 'Content preservation', *Phil. Rev.* **102**, 457–88

—— (1998), 'Computer proof, a priori knowledge, and other minds', *Phil. Perspectives* **12**, 1–37

Burgess, John (1998), 'Occam's razor and scientific method', in Schirn (1998), pp. 195–214

—— (2004), 'Mathematics and *Bleak House*', *Philosophia Mathematica* **12**, 18–36

—— (2005), *Fixing Frege*, Princeton University Press

—— , and Gideon Rosen (1997), *A Subject with no Object: Strategies for nominalistic interpretation of mathematics*, Oxford University Press

—— (2005), 'Nominalism reconsidered', in Shapiro (2005), pp. 515–35

Butterworth, Brian (1999), *The Mathematical Brain*, Macmillan

Cappelletti, M., A. Jansari, M. Kopelman, and B. Butterworth (forthcoming), 'A case of selective impairment of encyclopaedic numerical knowledge or '*when December 25th is no longer Christmas day, but '20+5' is still 25*', *Cortex*

—— , B. Butterworth, and M. D. Kopelman (2001), 'Spared numerical abilities in a case of semantic dementia', *Neuropsychologia* **39**, 1224–39

—— , H. L. Lee, and C. J. Price (2007), 'Dissociating the neural systems for conceptual processing of numerals and object names', *Journal of Cognitive Neuroscience*

—— , M. D. Kopelman, J. Morton, and B. Butterworth (2005), 'Dissociations in numerical abilities revealed by progressive cognitive decline in a patient with semantic dementia', *Cognitive Neuropsychology* **22**, 771–93

Carroll, J. B. (ed.) (1956), *Language, Thought, and Reality: Selected writings of Benjamin Lee Whorf*, MIT Press, Cambridge, MA

Chen, J. (1978), 'On the representation of a large even integer as the sum of a prime and the product of at most two primes', *Jl. Sci. Sinica* **21**(4), 421–30

Cipolotti, L., B. Butterworth, and G. Denes (1991), 'A specific deficit for numbers in a case of dense acalculia', *Brain* **114**, 2619–37

Colyvan, Mark (2001), *The Indispensability of Mathematics*, Oxford University Press

—— (2002), 'Mathematics and aesthetic considerations in science', *Mind* **111**, 69–74

—— (2006), 'Scientific realism and mathematical nominalism: A marriage made in hell', in C. Cheyne, and J. Worrall (eds), *Rationality and Reality: Conversations with Alan Musgrave*, Springer, Dordrecht, pp. 225–37

Cook, Roy (2002), 'The state of the economy: Neo-logicism and inflation', *Philosophia Mathematica* **10**, 43–66

Corfield, David (2005), *Towards a Philosophy of Real Mathematics*, Cambridge University Press

Crutch, S. J., and E. K. Warrington (2002), 'Preserved calculation skills in a case of semantic dementia', *Cortex* **38**, 389–99

Davies, Martin (2000), 'Externalism and armchair knowledge', in P. Boghossian, and C. Peacocke (eds), *New Essays on the A Priori*, Oxford University Press

Davis, Philip J., and Reuben Hersh (1981), *The Mathematical Experience*, Birkhäuser, Boston, MA

Dehaene, S. (2001), 'Précis of "The Number Sense"', *Mind and Language* **16**, 16–36

— (1997), *The Number Sense*, Oxford University Press

— , and L. Cohen (1995), 'Towards an anatomical and functional model of number processing', *Mathematical Cognition* **1**, 83–120

— , M. Piazza, P. Pinel, and L. Cohen (2003), 'Three parietal circuits for number processing', *Cognitive Neuropsychology* **20**, 487–506

— , E. Spelke, R. Stanescu, P. Pinel, and S. Tsivkin (1999), 'Sources of mathematical thinking: behavioral and brain-imaging evidence', *Science* **284**, 970–74

Delazer, M., and B. Butterworth (1997), 'A dissociation of number meanings', *Cognitive Neuropsychology* **14**, 613–36

Desboves, A. (1855), *Nouv. Ann. Math.* **14**, 293

Dummett, Michael (1978), 'Wang's paradox', in *Truth and other Enigmas*, Duckworth, pp. 248–68

— (1991), *Frege: Philosophy of Mathematics*, Duckworth

Echeverria, J. (1996), 'Empirical methods in mathematics. A case study: Goldbach's conjecture', in G. Munévar (ed.), *Spanish Studies in the Philosophy of Science*, Kluwer

Evans, Gareth (1982), *The Varieties of Reference*, Oxford University Press

Feferman, Solomon (1992), 'Why a little bit goes a long way: Logical foundations of scientifically applied mathematics', *PSA: Proceedings of the Biennial Meeting of the Philosophy of Science Assocation* **2**, 442–55

— (1998), *In the Light of Logic*, Oxford University Press

Feigenson, L., S. Carey, and M. Hauser (2002), 'The representations underlying infants' choice of more: object files versus analog magnitudes', *Psychological Science* **13**, 150–6

Field, Hartry (1980), *Science Without Numbers: A defence of nominalism*, Princeton University Press

— (1984), 'Is mathematical knowledge just logical knowledge?', *Philosophical Review* **93**, 509–52. Reprinted with a postscript in (Field 1989: pp. 79–124)

— (1989), *Realism, Mathematics and Modality*, Blackwell

— (1991), 'Metalogic and modality', *Philosophical Studies* **62**(1), 1–22

Fine, Kit (2002), *The Limits of Abstraction*, Oxford University Press

Flew, Anthony, and Stephen Priest (2002), *A Dictionary of Philosophy*, Macmillan

Forster, M., and E. Sober (1994), 'How to tell when simpler, more unified, or less ad hoc theories will provide more accurate predictions', *Brit. J. Phil. Sci.* **45**, 1–36

Frege, G. (1970), *Translations from the Philosophical Writings of Gottlob Frege*, Blackwell, Oxford

—— (1974), *The Foundations of Arithmetic*, Blackwell, Oxford

Fuson, K. C. (1992), 'Relationship between counting and cardinality from age 2 to 8', in J. Bideaud, C. Meljac, and J. P. Fisher (eds), *Pathways to Number: Children's Developing Numerical Abilities*, Erlbaum, Hillsdale, NJ

Garavaso, P. (2005), 'On Frege's alleged indispensability argument', *Philosophia Mathematica* **13**, 160–73

Gardner, Martin (1981), 'Is mathematics for real?', *New York Review of Books* **28**(13)

Gettier, Edmund L. (1963), 'Is justified true belief knowledge?', *Analysis* **23**, 121–23

Ginzburg, Lev R., and Mark Colyvan (2004), *Ecological Orbits: How planets move and populations grow*, Oxford University Press

Gödel, Kurt (1944), 'Russell's mathematical logic', in *The philosophy of Bertrand Russell*, North Western University Press, Chicago, IL, pp. 123–53

—— (1995), 'Some basic theorems on the foundations of mathematics and their philosophical implications', in *Collected Works*, Vol. III, Oxford University Press, pp. 304–23. The Gibbs Lecture delivered at Brown University.

Goodman, N., and W. V. Quine (1947), 'Steps towards a constructive nominalism', *J. Symb. Logic* **12**, 105–22

Gordon, P. (2004), 'Numerical cognition without words: Evidence from amazonia', *Science* **306**, 496–9

Guy, R. (1994), *Unsolved problems in Number Theory*, Springer, New York

Hacking, Ian (1983), *Representing and Intervening*, Cambridge University Press

Hale, Bob (1994), 'Dummett's critique of Wright's attempt to resuscitate Frege', *Philosophia Mathematica* **3**, 122–47

—— (2000*a*), 'Abstraction and set theory', *Notre Dame Journal of Formal Logic* **41**, 379–98

—— (2000*b*), 'Reals by abstraction', *Philosophia Mathematica* **8**, 100–23

Hale, Bob, and Crispin Wright (2000), 'Implicit definition and the a priori', in P. Boghossian, and C. Peacocke (eds), *New Essays on the A Priori*, Oxford University Press, pp. 286–319

—— (2001), *The Reason's Proper Study*, Oxford University Press

Hardy, G. H. (1929), 'Mathematical proof', *Mind* **38**(149), 1–25

Hart, W. D. (1977), 'Review of Steiner, Mathematical Knowledge', *Journal of Philosophy* **74**, 118–29

Hebb, D. O. (1949), *The Organization of Behavior*, John Wiley and Sons, New York

Heider, E. R., and D. Oliver (1972), 'The structure of the colour space in naming and memory for two languages', *Cognitive Psychology* **3**, 337–54

Hellman, G. (1999), 'Some ins and outs of indispensability: A modal-structural perspective', in A. Cantini, E. Casari, and P. Minari (eds), *Logic and Foundations of Mathematics*, Kluwer, Dordrecht, pp. 25–39

— (2005), 'Structuralism', in Shapiro (2005), pp. 536–62

Hellman, Geoffrey (1989), *Mathematics without Numbers*, Oxford University Press

Henschen, S. (1920), *Klinische und anatomische Beiträge zur Pathologie des Gehirns*, Vol. 5, Nordiska, Stockholm

Hersh, R. (ed.) (2005), *18 Unconventional Essays on the Nature of Mathematics*, Springer, Berlin

Hersh, Reuben (1997), *What is Mathematics, Really?*, Oxford University Press

Hespos, S. J. (2004), 'Language: Life without numbers', *Current Biology* **14**, R927–8

Hodges, J. R., K. Patterson, S. Oxbury, and E. Funnell (1992), 'Semantic dementia: Progressive fluent aphasia with temporal lobe atrophy', *Brain* **115**, 1783–1806

Jubien, M. (1989), 'On properties and property theory', in G. Chierchia, B. Partee, and R. Turner (eds), *Properties, Types and Meaning*, Vol. 1, Kluwer, pp. 159–75

Kitcher, Philip (1984), *The Nature of Mathematical Knowledge*, Oxford University Press

Koehler, O. (1951), 'The ability of birds to count', *Bulletin of Animal Behaviour* **9**, 41–5

Kreisel, Georg (1967), 'Informal rigour and completeness proofs', in I. Lakatos (ed.), *Problems in the Philosophy of Mathematics*, North-Holland, Amsterdam, pp. 138–71

Lakatos, Imre (1978), 'Introduction: Science and pseudoscience', in *The Methodology of Scientific Research Programmes*, Cambridge University Press

Lakoff, George, and Rafael E. Núñez (2000), *Where Mathematics Comes From*, Basic Books, New York

Laudan, L. (1981), 'A problem-solving approach to scientific progress', in I. Hacking (ed.), *Scientific Revolutions*, Oxford University Press, pp. 144–55

Leivant, Daniel (1994), 'Higher-order logic', in D. M. Gabbay, C. J. Hogger, J. A. Robinson, and J. Siekmann (eds), *Handbook of Logic in Artificial Intelligence and Logic Programming*, Vol. 2, Oxford University Press, pp. 230–321

Lemer, C., S. Dehaene, E. Spelke, and L. Cohen (2003), 'Approximate quantities and exact number words: dissociable systems', *Neuropsychologia* **41**, 1942–58

Leng, Mary (2002), 'What's wrong with indispensability? (Or, The case for recreational mathematics)', *Synthese* **131**, 395–417

— (2005*a*), 'Mathematical explanation', in C. Celluci, and D. Gillies (eds), *Mathematical Reasoning and Heuristics*, King's College Publications, London, pp. 167–89

— (2005*b*), 'Platonism and anti-platonism: Why worry?', *International Studies in the Philosophy of Science* **19**, 65–84

— (2008), *Mathematics and Reality*, Oxford University Press

Levinson, S. C. (1996), 'Relativity in spatial conception and description', in J. J. Gumperz, and S. C. Levinson (eds), *Rethinking Linguistic Relativity*, Cambridge University Press

Lewis, David (1986), *On the Plurality of Worlds*, Blackwell, Oxford

—— (1991), *Parts of Classes*, Blackwell, Oxford

—— (1993), 'Mathematics is megethology', *Philosophia Mathematica* **3**, 3–23

Lyon, Adrian, and Mark Colyvan (2008), 'The explanatory power of phase spaces', *Philosophia Mathematica* **16**

Maddy, Penelope (1990), *Realism in Mathematics*, Oxford University Press

—— (1992), 'Indispensability and practice', *J. Phil.* **89**, 275–89

—— (1995), 'Naturalism and ontology', *Philosophia Mathematica* **3**, 248–70

—— (1997), *Naturalism in Mathematics*, Oxford University Press

Malinas, Gary, and John Bigelow (2004), 'Simpson's paradox', in E. Zalta (ed.), *The Stanford Encyclopaedia of Philosophy* **URL:** http://plato.stanford.edu/archives/spr2004/entries/paradox-simpson/

Matsuzawa, T. (1985), 'Use of numbers by a chimpanzee', *Nature* **315**, 57–9

McCulloch, Warren (1965), *Embodiments of Mind*, MIT Press, Cambridge, MA

Melia, J. (2002), 'Reply to Colyvan', *Mind* **111**, 75–9

Mill, J. S. (1947), *A System of Logic*, Longmans, Green & Co.

Moyer, R. S., and T. K. Landauer (1967), 'Time required for judgements of numerical inequality', *Nature* **21**, 1519–20

Nieder, A. (2005), 'Counting on neurons: the neurobiology of numerical competence', *Nature Reviews Neuroscience* **6**, 177–90

Nolan, Daniel (1997), 'Impossible worlds: A modest approach', *Notre Dame Journal of Formal Logic* **38**, 535–72

O'Bryant, K. (n.d.), 'Goldbach's conjecture' **URL:** http://www.math.ucsd.edu/~kobryant

Parsons, Charles (1983), 'Quine on the philosophy of mathematics', in *Mathematics in Philosophy*, Cornell University Press, Ithaca, NY, pp. 176–205

Paseau, Alexander (2005), 'Naturalism in mathematics and the authority of philosophy', *Brit. J. Phil. Sci.* **56**, 399–418

Piaget, Jean (1937), *The Construction of Reality in the Child*, 1954 edn, Basic Books, New York

Pica, P., C. Lemer, V. Izard, and S. Dehaene (2004), 'Exact and approximate arithmetic in an Amazonian indigene group', *Science* **306**(5695), 499–503

Pincock, C. (2004*a*), 'A new perspective on the problem of applying mathematics', *Philosophia Mathematica* **12**, 135–61

—— (2004*b*), 'A revealing flaw in Colyvan's indispensability argument', *Philosophy of Science* **71**, 61–79

Polk, T. A., C. L. Reed, J. M. Keenan, P. Hogarth, and C. A. Anderson (2001), 'A dissociation between symbolic number knowledge and analogue magnitude information', *Brain and Cognition* **47**, 545–63

Polya, G. (1954), *Mathematics and Plausible Reasoning*, Vol. 1, Princeton University Press, Princeton, NJ

Potter, Michael (2000), *Reason's Nearest Kin: Philosophies of Arithmetic from Kant to Carnap*, Oxford University Press

— (2001), 'Was Gödel a Gödelian platonist?', *Philosophia Mathematica* **9**, 331–46

Potter, Michael, and T. J. Smiley (2001), 'Abstraction by recarving', *Proceedings of the Aristotelian Society* **101**, 327–38

Putnam, H. (1983), 'Mathematics without foundations', in P. Benacerraf, and H. Putnam (eds), *Philosophy of Mathematics: Selected Readings*, Cambridge University Press, pp. 295–311

Putnam, Hilary (1971), *Philosophy of Logic*, Harper, New York

— (1975), 'The meaning of 'meaning'', in *Mind, Language and Reality*, Cambridge University Press, pp. 215–71

— (1979), *Mathematics, Matter and Method*, Vol. 1 of *Philosophical Papers*, 2nd edn, Cambridge University Press

Quine, W. V. (1953), 'Two dogmas of empiricism', in *From a Logical Point of View* (Quine 1953), pp. 20–46

— (1960), *Word and Object*, MIT Press, Cambridge, MA

— (1970), *Philosophy of Logic*, Harvard University Press, Cambridge, MA

— (1974), *The Roots of Reference*, Open Court, La Salle, IL

— (1981*a*), 'Success and limits of mathematization', in *Theories and Things*, Harvard University Press, Cambridge, MA, pp. 148–55

— (1981*b*), *Theories and Things*, Harvard University Press

— (1986), 'Reply to Charles Parsons', in *The Philosophy of W. V. Quine*, Open Court, pp. 396–403

— (1995), *From Stimulus to Science*, Harvard University Press, Cambridge, MA

Quine, Willard Van Orman (1953), *From a Logical Point of View*, Harvard University Press, Cambridge, MA

Ramaré, O. (1995), 'On Schnirelman's constant', *Ann. Scuola Norm. Sup. Pisa Cl. Sci.* **22**(4), 645–706

Rayo, Agustin, and Stephen Yablo (2001), 'Nominalism through de-nominalisation', *Nous* **35**, 74–92

Remond-Besuchet, C., M. P. Noel, X. Seron, M. Thioux, M. Brun, and X. Aspe (1999), 'Selective preservation of exceptional arithmetical knowledge in a demented patient', *Mathematical Cognition* **5**, 65–91

Resnik, M. D. (1997), *Mathematics as a Science of Patterns*, Oxford University Press

Resnik, Michael (2005), 'Quine and the web of belief', in Shapiro (2005), pp. 412–36

Rosen, G. (1992), 'Remarks on Modern Nominalism', PhD thesis, Princeton

Rosen, Gideon (1994), 'What is constructive empiricism?', *Philosophical Studies* **74**, 143–78

Rossberg, Marcus (n.d.), Inferentialism and conservativeness

Rossor, M. N., E. K. Warrington, and L. Cipolotti (1995), 'The isolation of calculation skills', *Neurology, Neurosurgery and Psychiatry* **242**, 78–81

Russell, Bertrand (1945), *The History of Western Philosophy*, Simon & Schuster, New York

Salmon, W. (1996), 'Rationality and objectivity in science *or* Tom Kuhn meets Tom Bayes', in D. Papineau (ed.), *The Philosophy of Science*, Oxford University Press, pp. 256–89

Schirn, M. (ed.) (1998), *Philosophy of Mathematics Today*, Oxford University Press

Shalkowski, Scott A. (1994), 'The ontological ground of the alethic modality', *Philosophical Review* **103**(4), 669–88

Shallice, T. (1988), *From Neuropsychology to Mental Structure*, Cambridge University Press

Shapiro, Stewart (1993), 'Modality and ontology', *Mind* **102**, 455–81

—— (1997), *Philosophy of Mathematics: Structure and Ontology*, Oxford University Press

—— (2000), 'Frege meets Dedekind: A neologicist treatment of real analysis', *Notre Dame Journal of Formal Logic* **41**, 335–64

—— (2003), 'Prolegomena to any future neo-logicist set theory: Abstraction and indefinite extensibility', *Brit. J. Phil. Sci.* **54**, 59–91

Shapiro, Stewart, and Alan Weir (1999), 'New V, ZF and abstraction', *Philosophia Mathematica* **7**, 291–321

Shapiro, Stewart (ed.) (2005), *The Oxford Handbook of Philosophy of Mathematics and Logic*, Oxford University Press

Simon, O., J. F. Mangin, L. Cohen, M. Bruandet, Pinel P., F. Hennel, J. B. Poline, D. L. Bihan, and S. Dehaene (2002), 'Topographical arrangement of hand, eye, calculation, and language related areas in the human intraparietal sulcus', *Neuron* **33**, 475–87

Slonimsky, N., and Robert Bonotto (2002), *Slonimsky's Book of Musical Anecdotes*, Routledge

Snowden, J. S., D. Neary, and D. M. A. Mann (1996), *Frontotemporal Lobar Degeneration: Frontotemporal Dementia, Progressive Aphasia, Semantic Dementia*, Churchill Livingstone, New York

Sober, E. (1993*a*), 'Mathematics and indispensability', *Phil. Rev.* **102**, 35–57

—— (1993*b*), *Philosophy of Biology*, Oxford University Press

—— (1994*a*), 'Contrastive empiricism', in *From a Biological Point of View* (Sober 1994*b*), pp. 114–35

— (1994*b*), *From a Biological Point of View*, Cambridge University Press

— (1994*c*), 'Let's razor Ockham's razor', in *From a Biological Point of View* (Sober 1994*b*), pp. 136–57

— (1996), 'Parsimony and predictive equivalence', *Erkenntnis* **44**, 167–97

Spelke, E. S., K. Breinlinger, J. Macomber, and K. Jacobson (1992), 'Origins of knowledge', *Psychological Review* **99**, 605–32

Spelke, E. S., and S. Tsivkin (2001), 'Language and number: A bilingual training study', *Cognition* **78**, 45–88

Steiner, Mark (1975), *Mathematical Knowledge*, Oxford University Press

— (2005), 'Mathematics—Application and applicability', in Shapiro (2005), pp. 625–50

te Riele, H. (1987), 'On the sign of the difference $\pi(x) - \text{Li}(x)$', *Math. Comput.* **48**, 323–8

Thioux, M., A. Pillon, D. Samson, M.P. de Partz, M.P. Noël, and X. Seron (1998), 'The isolation of numerals at the semantic level', *Neurocase* **4**, 371–89

van Bendegem, J. (1998), 'What, if anything, is an experiment in mathematics', in D. Anapolitanos et al. (eds), *Philosophy and the Many Faces of Science*, Rowman and Littlefield

Wagner, S. (1996), 'Prospects for platonism', in A. Morton, and S. Stich (eds), *Benacerraf and his Critics*, Blackwell, Oxford, pp. 73–99

Wanga, S. H., R. Baillargeon, and L. Brueckner (2004), 'Young infants' reasoning about hidden objects: Evidence from violation-of-expectation tasks with test trials only', *Cognition* **93**, 167–98

Weir, Alan (2005), 'Naturalism reconsidered', in Shapiro (2005), pp. 460–82

Wigner, E. P. (1960), 'The unreasonable effectiveness of mathematics in the natural sciences', *Communications on Pure and Applied Mathematics* **13**, 1–14

Williamson, Timothy (2003), 'Understanding and inference', *Proc. Arist. Soc. Supp. Vol.* **77**, 249–73

— (2006), 'Conceptual truth', *Proc. Arist. Soc.* **80**, 1–41

Wittgenstein, Ludwig (1922; corrected edn. 1933), *Tractatus Logico-Philosophicus*, Kegan Paul & Trubner

— (1956), *Remarks on the Foundations of Mathematics*, 1978, 3rd edn, Blackwell, Oxford

Wright, Crispin (1983), *Frege's Conception of Numbers as Objects*, Aberdeen University Press

— (1998), 'On the (harmless) impredicativity of Hume's Principle', in Schirn (1998), pp. 339–68

Wynn, K. (1992), 'Addition and subtraction by human infants', *Nature* **358**, 749–50

Index